Physics by Computer

Springer-Verlag Berlin Heidelberg GmbH

Wolfgang Kinzel   Georg Reents

Physics by Computer

Programming Physical Problems
Using Mathematica® and C

Translated by Martin Clajus and Beverly Freeland-Clajus

 Springer

Prof. Dr. Wolfgang Kinzel
Priv.-Doz. Dr. Georg Reents
Institut für Theoretische Physik
Universität Würzburg
Am Hubland
D-97074 Würzburg
e-mail:
kinzel@physik.uni-wuerzburg.de
reents@physik.uni-wuerzburg.de

Dr. Martin Clajus
Dr. Beverly Freeland-Clajus
Dept. of Physics & Astronomy
UCLA
Box 951547
Los Angeles, CA 90095-1547, USA
e-mail:
clajus@physics.ucla.edu

The cover picture shows a space-time diagram of the probability of presence of a quantum particle in a square-well potential.

This American edition has been revised by the authors. The book was originally published in German: W. Kinzel, G. Reents: *Physik per Computer,* © Spektrum Akademischer Verlag, Heidelberg 1996

ISBN 978-3-642-46841-4 ISBN 978-3-642-46839-1 (eBook)
DOI 10.1007/978-3-642-46839-1

CIP data applied for
Die Deutsche Bibliothek – CIP-Einheitsaufnahme

Cover design: Künkel + Lopka, Heidelberg
Typesetting: Camera-ready copy from the translator using a Springer TEX macro package
SPIN: 10546058   56/3144-5 4 3 2 1 0 – Printed on acid-free paper

Preface

Nowadays the computer is an important tool in physics. The acquisition and analysis of extensive experimental data and the control of complex experiments are hardly imaginable without the use of computers. In theoretical physics the computer has turned from a mere calculator to a comprehensive tool. Graphical displays, numerical and algebraic solutions of equations, and extensive simulations of microscopic models have become important methods for the exploration of the laws of physics.

The computer, however, is not just a tool, it also offers new perspectives and opens new areas of research. Until recently physicists generally described nature with differential equations; nowadays discrete algorithms are also used. For some apparently simple physical models there are only numerical answers so far. We know universal laws that any high school student can reproduce on a pocket calculator, for which there is, however, no analytical theory (yet?). In addition to this, the computer opens up new fields to physics: neural networks, combinatorial optimization, biological evolution, formation of fractal structures, and self-organized criticality are just some of the topics from the growing field of complex systems.

Almost every advanced undergraduate or graduate physics student uses a computer at one time or another. Nonetheless, computer training and the use of computers in teaching are still by no means the expected norm, but rather the exception. The literature in this field is correspondingly sparse. The goal of our textbook is to contribute to filling this gap.

This book evolved out of lectures at the University of Würzburg, Germany, for physics majors after their fourth semester – those having completed the introductory coursework in theoretical physics. It is conceived as a textbook in computational physics but may also serve as a supplement to the traditional physics classes in which the possibilities of computer use have so far been underutilized. We would like to show the reader how to solve physics problems using the computer. Experience with computers and computer languages is helpful, but not necessary, for we want to present an introduction and explain the first steps with computers and algorithms. This book does not contain many details about numerical mathematics, it does not offer a course on programming languages, nor does it teach advanced methods of computer-oriented theoretical physics. It does, however, introduce numerous physics problems, some of which are at the cutting edge of research, and tries to solve them with simple algorithms.

One goal is to encourage our readers to do their own programming. Although a CD-ROM with finished programs is enclosed with the book, they are not meant as user-friendly experimental environments. We hope that instead they can be taken as a starting-point, and we encourage our readers to modify them, or better yet to rewrite them more efficiently. Exercises accompany every section of this introductory book.

We have received suggestions from many colleagues and students, to whom we wish to express our thanks. We would like especially to mention M. Biehl, H. Dietz, A. Engel, A. Freking, Th. Hahn, W. Hanke, G. Hildebrand, A. Jung, A. Karch, U. Krey, B. Lopez, J. Nestler, M. Opper, M. Schreckenberg, and D. Stauffer. Special thanks go to the following three people: Martin Lüders developed the program package *Xgraphics*, Martin Schröder wrote the section on Unix, and Ursula Eitelwein typed the manuscript of this book in LaTeX. Finally we would like to thank Martin Clajus for valuable suggestions in the course of translating this textbook into English.

Würzburg   *Wolfgang Kinzel*
July, 1997   *Georg Reents*

Contents

Introduction

This text is intended to introduce the reader to the solution of physics problems with the computer. The examples have been kept as simple as possible, in order to minimize the effort of expressing them algebraically, numerically, and graphically. The use of computers is learned mainly in the course of one's own work – readiness for continuous experimenting, playful curiosity, and the desire to research are important prerequisites for obtaining the maximum benefit from this book.

Almost every section contains three parts with the following titles: Physics, Algorithm, and Results. In the physics part, the problem is formulated and the model is introduced. In the middle (algorithm) part, the computer program which is supposed to solve the problems from the first part is described. Only the essential commands are explained so that the readers can try to complete the programs themselves. Alternatively, the complete computer code can be found in the appendix for further use; the source and executable codes are available on the enclosed CD-ROM. In the results part, some results of the computer calculations are illustrated, for the most part in the form of graphics.

We have consciously chosen to handle a large number of examples; some are taken from current research. Therefore the presentation of the theoretical foundations can only be seen as a first introduction. Many sections could be expanded into entire courses; we can only refer the reader to additional literature. In addition, we have cited some textbooks or selected original literature for each problem. The choice of these citations is, of course, subjective and no claim of completeness is made. At the end of every section, there is an exercise.

As programming languages we have chosen *Mathematica* and C. Both are modern languages that are particularly suited for today's workstation computers (e.g., RISC workstations running under Unix). This choice is subjective and in no way necessary for working through the problems presented in this text. The language Fortran, whose new version Fortran90 has taken over elements of C, is often used for CPU-intensive computer simulations. MAPLE is also a very popular tool for computer algebra. All of our programs run not only on fast workstations, but also on a PC operating under DOS or Linux.

Minimum requirement for the PC programs is a 386 computer with co-processor and VGA card. We have developed the *Mathematica* examples with version 2.2 of the program, but they also run under version 3.0. Turbo C (2.0) by Borland was used for the C programs on the PC. One can easily develop programs and generate graphics with this programming environment.

There are also workstation versions of all programs; they have been tested on a Hewlett Packard HP 9000 Series 700 workstation. Unfortunately it is much more difficult to produce graphics there than on a PC. In order to draw just a line or a circle on the screen, displays have to be set, windows generated, color palettes and, if necessary, events (mouse or keyboard commands) have to be defined, and the contents of the display memory have to be written to the monitor. To do so, one has to learn complicated new languages like X11, Motif, or PHIGS. In order to make this easier, Martin Lüders has written the program package *Xgraphics*, which is based on the X11 library and will generate on, hopefully, any workstation monitor elementary graphics with relatively few commands. This package is included on the enclosed CD-ROM. In addition, any interested person can download it via the Internet from our FTP server and compile it. Versions of all C programs that use this graphics package are included as well.

All C programs in this book are described in the PC version. However, the figures were produced with *Xgraphics*. For the most part we have shown the windows as they appear on the monitor.

Most C programs can be controlled interactively during their execution. The versions for the PC as well as those for the workstation allow the user to change some parameters or algorithms or end the program via keyboard inputs. On the workstation, the execution can also be controlled with the mouse.

We want to emphasize again that our hardware and software equipment are by no means necessary for working through the physics problems. Every problem can also be solved with various other programming languages. We hope that the readers of this book will soon be in the position to formulate and work through the models themselves, in their favorite language.

1. Functions in *Mathematica*

At the lowest hierarchical level of programming languages, any computer processes small packets of on–off data (bits) step by step. Every packet gives instructions to electronic switches that cause the results of elementary mathematical operations to be written to memory and new data to be read. This level of processing elementary instructions step by step, however, is hardly suited to the formulation of more involved problems by humans.

This is why complex algorithms should be structured. Correspondingly, high-level programming languages consist of functions or modules. In order to keep a program comprehensible, one should use modules that are organized from coarse to fine. As a rule, a good programming style will involve multiple levels of modules; this has the advantage of simplifying the modification of such programs by making it more likely that only individual functions will have to be exchanged.

Mathematica is a programming language containing a large number of functions that are easily called, however, since most parameters have default values and do not necessarily have to be specified. Indeed, every command in *Mathematica* can be regarded as a function that is executed immediately and is relevant for all subsequent and sometimes even for previous commands. In this sense, *Mathematica* is located right at the top of the hierarchy of programming languages.

By comparison, C, Fortran, Pascal, and Basic, although they, too, are higher programming languages offering numerical and graphical functions (after all, *Mathematica* is written in C), require much more effort than *Mathematica* in order to, say, generate a graph of a Bessel function in the complex plane.

This chapter is intended to introduce the reader to the use of *Mathematica*. With simple examples, the following applications are demonstrated: elliptic integrals, series expansion, Fourier transformation, χ^2 distributions, 3D graphics, multiplication and integration of vectors, determining the zeros of nonlinear equations, and linear optimization.

1.1 Function versus Procedure

In order to illustrate the difference between processing commands step by
step, a method for which we want to use the term "procedure", and calling a predefined function, we will use a simple example: the mean value of
10 000 random numbers, distributed uniformly in the interval $[0, 1]$, is to be
determined. In *Mathematica*, the data are generated as follows:

```
dataset = Table[Random[],{10000}]
```

The result is a list of 10 000 numerical values.

In order to determine the mean value, we have to add up all these numbers
and divide the sum by the number of values. We can program this step by
step and thus write a procedure we will call **average**. The iteration can be
written as a Do or For loop; the function Block shields the variables sum and
average from external definitions. This procedure might look like this:

```
average[data_,length_]:=
              Block[ {sum, average},
              sum = 0. ;
              Do[sum = sum + data[[i]], {i,length}];
              average = sum/length ]
```

Here, step by step, a number from the data set **data** is added to the variable
sum, and then the result is divided by the length of **data**, which has to be
supplied to the procedure. The solution of the same problem looks quite
different when a modular structure is used:

```
average[data_] := Apply[Plus, data]/Length[data]
```

In this case we use functions that take advantage of the structure of *Mathematica*. The argument **data** is entered as a list of numbers. **Apply** replaces the
header List of the list **data** by the header Plus, which causes all elements
of **data** to be added up. **Length** determines the list's length.

Note that **average** appears three times: as a local variable and as the
name of two functions, one of which has two arguments, the other just one.
Mathematica recognizes the difference and uses the correct expression in each
case.

The reader will find an introduction to *Mathematica* in Appendix A; nevertheless, we want to point out some features of this language here: the underscore character in the argument **data_** indicates that it can be replaced
by any symbol which will be transferred to the function. An assignment with
:= reevaluates the right-hand side each time the function is called, whereas
= evaluates it only once. For example, r := Random[] will result in a new
value each time r is called, whereas repeated calls of r will return the same
value for r = Random[]. Indices of lists (arrays, vectors, matrices, tensors,
etc.) are represented by double brackets [[i]].

In order to illustrate the difference between *Mathematica* and the programming language C, we show our sample procedure again, this time in C. We increase the number of array elements to be averaged from 10 000 to one million so that we obtain run times of the order of seconds:

```
#include <stdlib.h>
#include <time.h>

#define LIMIT 1000000

main()
{
 double average(double *,long), dataset[LIMIT];
 clock_t start,end;
 long i;
 for (i=0;i<LIMIT;i++) dataset[i]=(double)rand()/RAND_MAX;
 start=clock();
 printf(" average = %f\n",average(dataset,LIMIT));
 end=clock();
 printf("time= %lf\n",(double)(end-start)/CLOCKS_PER_SEC);
}

double average( double *data,long n )
{
 double sum=0. ;
 long i;
 for(i=0;i<n;i++) sum=sum+data[i] ;
 return sum/n;
}
```

Note the differences between *Mathematica* and C. In C, everything has to be declared: library routines are declared in the header files <name.h>, self-defined variables and functions by int, char, float, double, etc. C requires a sequence of instructions with a clearly defined main routine and functions, whereas *Mathematica* always attempts to evaluate all symbols by using existing definitions. C is machine oriented which has advantages and disadvantages. For example, what is transferred to the function average is the memory location of the array dataset; more precisely, dataset is a pointer to (= the address of) the array, and its individual elements are referenced by dataset[0], ..., dataset[999999]. Pointers are declared via a *, so double *data means that data contains the address of a variable of type double. Warning: dataset[1000000] is some undefined entity; referencing it will not cause the computer to report an error but will usually produce completely incomprehensible results.

Unfortunately, a comma, semicolon, and various kinds of parentheses have completely different meanings in C and *Mathematica*. But why should life be easy for computer users?

Now what is the result of the three versions of our averaging program? If one enters average[dataset] in *Mathematica*, a PC will return a value close to 0.5 after about ten seconds. By comparison, the version `average[dataset, 10000]`, modeled after the C program, takes three to four times longer to yield, of course, the same result. In each case, the workstation calculates these results about ten times faster than the PC.

The C program, however, is compiled first, for example by the command `gcc -o sum sum.c -lm`, and then called by its name `sum`. Now it takes only 8.5 seconds on the PC and 0.56 seconds on the HP to obtain the result, even though the number of steps has been increased by a factor of 100. This demonstrates a significant disadvantage of *Mathematica*: for numerical calculations, it is extremely slow compared to C or similar languages.

Literature

Kernighan B.W., Ritchie D.M. (1988) The C Programming Language. Prentice Hall, Englewood Cliffs, NJ

Wolfram S. (1996) The Mathematica Book, 3rd ed. Wolfram Media, Champaign, IL, and Cambridge University Press, Cambridge

1.2 The Nonlinear Pendulum

The mathematical pendulum is a standard example in any physics course on classical mechanics. The linear approximation of this problem is suitable for a high school physics course. The exact solution is expressed by an elliptic integral, but only a computer affords us the opportunity to visualize and thoroughly analyze these nonelementary functions. Therefore, we want to place this standard example of theoretical physics at the beginning of our textbook.

Physics

Let us start with the physical model: a pointlike mass m suspended from a massless, rigid string of length l swings in the earth's gravitational field. Let $\varphi(t)$ be the angle of displacement from the pendulum's equilibrium position at time t; friction is neglected. According to the laws of mechanics, the energy E of the pendulum is constant, so

$$E = \frac{m}{2} l^2 \dot{\varphi}^2 - m g l \cos\varphi = -m g l \cos\varphi_0 , \tag{1.1}$$

where φ_0 denotes the maximum displacement angle with $\dot{\varphi}_0 = 0$, and $\dot{\varphi} = d\varphi/dt$ is the angular velocity. From this equation, one can obtain $\varphi(t)$ by using a little trick: one separates t and φ by solving (1.1) first for $(d\varphi/dt)^2$ and then for the differential dt. For the half period in which φ increases with t, one obtains

$$dt = \sqrt{\frac{l}{2g}} \, \frac{d\varphi}{\sqrt{\cos\varphi - \cos\varphi_0}} . \tag{1.2}$$

For a suitable choice of $t = 0$ ($\varphi(0) = 0$, $\dot{\varphi}(0) > 0$), integration of this equation yields

$$t(\varphi) = \sqrt{\frac{l}{2g}} \int_0^\varphi \frac{d\varphi'}{\sqrt{\cos \varphi' - \cos \varphi_0}} \ . \tag{1.3}$$

Obviously, the period T is four times the time the pendulum needs to reach its maximum angle φ_0:

$$T = 4\, t(\varphi_0) \ . \tag{1.4}$$

Since the integrand of (1.3) diverges as $\varphi' \to \varphi_0$, it makes sense to use the substitution

$$\sin \psi = \frac{\sin(\varphi/2)}{\sin(\varphi_0/2)} \ , \tag{1.5}$$

which results in the following integral:

$$t(\varphi) = \sqrt{\frac{l}{g}} \int_0^\psi \frac{d\psi'}{\sqrt{1 - \sin^2(\varphi_0/2)\sin^2 \psi'}} = \sqrt{\frac{l}{g}}\, F\left(\psi, \sin \frac{\varphi_0}{2}\right) \ . \tag{1.6}$$

Now, the integrand diverges only for an amplitude $\varphi_0 = \pi$. The function F is called an elliptic integral of the first kind. According to (1.5), $\psi = \pi/2$ for $\varphi = \varphi_0$; therefore, the period T is

$$T = 4\sqrt{\frac{l}{g}}\, F\left(\frac{\pi}{2}, \sin \frac{\varphi_0}{2}\right) = 4\sqrt{\frac{l}{g}}\, K\left(\sin \frac{\varphi_0}{2}\right) \ . \tag{1.7}$$

The function K is called a complete elliptic integral of the first kind. Both F and K are available in *Mathematica*.

For small values of the displacement φ, the potential energy is approximately quadratic. As is well known, the force is a linear function of φ in this case, and the solution of the equation of motion is a sinusoidal oscillation with a period $T = 2\pi \sqrt{l/g}$. For larger amplitudes φ_0, the force becomes a nonlinear function of φ, and T depends on φ_0.

We want to study the effect of the nonlinearity. To be able to compare different curves for $\varphi(t)$, we normalize φ with φ_0 and t with T. In addition, we look at the phase-space diagram $\dot{\varphi}$ versus φ for various energies E. The generation of higher harmonics with increasing nonlinearity is visualized by a Fourier transformation. Finally we want to expand T in terms of the quantity $\sin^2(\varphi_0/2)$, which is a measure of the nonlinearity.

Algorithm

The functions $K(k)$ and $F(\psi, k)$ are available in *Mathematica* and are called by the commands

```
EllipticK[k^2]
```

and

```
EllipticF[psi, k^2]
```

One has to be careful about the arguments since, unfortunately, there are various conventions for using them. Thus, we have $K(k) =$ `EllipticK[k^2]` and correspondingly for the other elliptic integrals. According to (1.7), the period is simply obtained by

```
T[phi0_] = 4 EllipticK[Sin[phi0/2]^2]
```

where we have set $\sqrt{l/g} = 1$. The command `Plot` draws the period as a function of φ_0:

```
Plot[ T[phi0], {phi0, 0, Pi}, PlotRange -> {0,30} ]
```

Since T diverges as $\varphi_0 \to \pi$, one should use `PlotRange` and ignore *Mathematica*'s warning about $T(\pi)$.

The function $t(\varphi)$ is obtained from `EllipticF` according to (1.6). We, however, want to calculate the inverse, $\varphi(t)$. These functions are available as well; we have (using again $\sqrt{l/g} = 1$)

```
psi[t_,phi0_]=JacobiAmplitude[t, Sin[phi0/2]^2]
```

and

```
sinepsi[t_,phi0_]=JacobiSN[t, Sin[phi0/2]^2]
```

Now we replace the variable ψ with φ according to (1.5), normalize φ with φ_0 and t with T (defining $x = t/T$), and finally obtain the normalized function

```
phinorm[x_,phi0_]:=
   2 ArcSin[Sin[phi0/2] sinepsi[x T[phi0], phi0]]/phi0
```

Moreover, the function `JacobiSN[t, Sin[phi0/2]^2]` has the correct symmetry, so it provides the solution not only for $0 \leq x \leq 1/4$, but for all x. We want to have a look at the function $\varphi_{norm}(x) = \varphi(x)/\varphi_0$ for various values of the amplitude φ_0. To this end, we generate an array of five φ_0 values, `phi0[1]` $=$ `N[0.1 Pi]`, ..., `phi0[5]` $=$ `N[0.999 Pi]` and form a list of functions

```
flist = Table[phinorm[x,phi0[i]],{i,5}]
```

We can now draw this list using `Plot`, after first applying the command `Evaluate` to it.

In order to study the generation of higher harmonics with increasing nonlinearity φ_0, we expand $\varphi(t)$ into a Fourier series. This is most easily accomplished by using the built-in command `Fourier`; to do this, however, one has to generate a list of discrete values $\varphi(t_i), i = 1, \ldots, N$ (see next section). Then one obtains the absolute values of the complex Fourier coefficients as a list:

```
foulist = Abs[Fourier[list]]
```

If, in (1.1), the energy E is kept constant, one obtains curves in the $(\dot{\varphi}, \varphi)$ plane, the so-called phase space. Plotting these curves is surprisingly simple in *Mathematica*, by calling ContourPlot for the function $E(\dot{\varphi}, \varphi)$. The option Contours->{E1, E2, ..., En} can be used to have n contours plotted for various values of E.

Finally, a remark concerning the series expansion of $T(\varphi_0)$. The new version (3.0) of *Mathematica* allows for a direct Taylor expansion of (1.7) by using the command Series, whereas older versions will output only formal derivatives of of EllipticK.

A possible way to handle this problem is to start with the integral (1.6) for $\psi = \pi/2$. The integrand is

```
f = 1/Sqrt[1 - m Sin[psi]^2]
```

with $m = \sin^2(\varphi_0/2)$. f can be expanded in terms of m, e.g., up to 10th order in m about 0,

```
g = Series [f, {m,0,10}]
```

Then, every single term can be integrated over ψ by using Integrate. Finally, the value $\sin^2(\varphi_0/2)$ is substituted back for m via /. m -> Sin[phi0/2]^2.

Results

We have investigated the effect of the amplitude φ_0, which is a measure of the nonlinearity, on the motion of the pendulum. Figure 1.1 shows the period T as a function of the amplitude φ_0. It takes the pendulum an infinite amount of time to come to rest at the highest apex ($\varphi_0 = \pi$). We can see that at first, with increasing φ_0, T does not deviate greatly from the value for the harmonic pendulum, $T = 2\pi$. Only for amplitudes above $90° = \pi/2$ is T significantly larger. The influence of φ_0 on the curve $\varphi(t)$ is evident in Fig. 1.2, where the ratio φ/φ_0 is plotted as a function of t/T in order to compare different

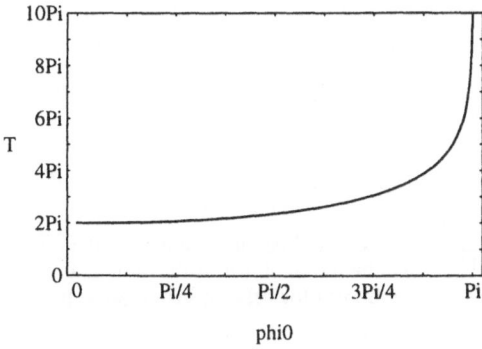

Fig. 1.1. The period T as a function of the amplitude φ_0

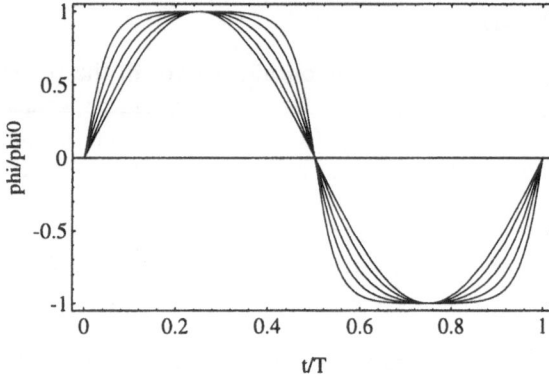

Fig. 1.2. Normalized displacement $\varphi_{\mathrm{norm}}(t/T)$ of the pendulum for various amplitudes $\varphi_0 = \pi/10$, $4\pi/5$, $19\pi/20$, $99\pi/100$, and $999\pi/1000$ (from the inside out)

curves (note: T diverges as $\varphi_0 \to \pi$). For small φ_0, we have a sinusoidal oscillation. For increasing φ_0, the time the pendulum spends near the apex $\pm\varphi_0$ becomes larger and larger; consequently, $\varphi(t)$ becomes flatter and flatter and turns into a step function for $\varphi_0 = \pi$.

If one decomposes $\varphi(t)$ into harmonic oscillations $b_s \exp(-i\omega_s t)$, then, because of the periodicity $\varphi(t) = \varphi(t + T)$, the frequencies ω_s are integer multiples of $2\pi/T$. The discrete Fourier transformation available in *Mathematica* takes the frequencies $\{\omega_s = (2\pi/T)(s-1),\ s = 1, \ldots, N\}$ at N data points $\{\varphi(t_r),\ r = 1, \ldots, N\}$ and calculates the coefficients according to

$$ b_s = \frac{1}{\sqrt{N}} \sum_{r=1}^{N} \varphi(t_r) \exp\left(2\pi i \frac{(s-1)(r-1)}{N}\right) . $$

The symmetry $\varphi(T/2 + t) = -\varphi(t)$ means that the coefficients b_s vanish for all odd s. The first nonvanishing coefficient, b_2, is the amplitude of the fundamental harmonic $\omega_2 = 2\pi/T$. The other coefficients $b_4, b_6, b_8, \ldots$ yield the amplitudes of the higher harmonics $3\omega_2, 5\omega_2, 7\omega_2, \ldots$. For $\varphi_0 = 0.999\pi$, the expression `foulist` contains the absolute values of the coefficients b_s, which are displayed via the command `ListPlot` in Fig. 1.3.

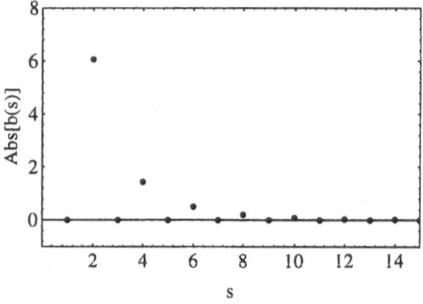

Fig. 1.3. Absolute values of the Fourier coefficients of the anharmonic oscillation $\varphi(t)$ for an amplitude $\varphi_0 = 0.999\pi$

The alternation between displacement φ and angular velocity $\dot{\varphi}$ can be illustrated in $(\varphi, \dot{\varphi})$ phase space; this eliminates the time t. Figure 1.4 shows such phase-space curves for different energies E. For small φ_0, one obtains a circle that becomes deformed as the energy increases. For $E > mgl$, the pendulum overshoots its apex; it moves in only one direction and has an angular velocity $\dot{\varphi} \neq 0$ even at the apex.

By using the command

```
tseries = 4 Integrate[g, {psi, 0, Pi/2}] /. m -> Sin[phi0/2]^2
```

one obtains the symbolic expansion of T in terms of $\sin^2(\varphi_0/2)$. *Mathematica* produces the following somewhat unwieldy output on the monitor:

```
               phi0 2            phi0 4                 phi0 6
        Pi Sin[----]      9 Pi Sin[----]        25 Pi Sin[----]
               2                 2                      2
2 Pi + -------------  +  ----------------  +  ------------------  +
              2                 32                   128


         phi0 8              phi0 10                  phi0 12
 1225 Pi Sin[----]   3969 Pi Sin[----]      53361 Pi Sin[----]
             2                 2                      2
> ------------------ + -------------------- + -------------------- +
        8192                 32768                 524288


          phi0 14               phi0 16
 184041 Pi Sin[----]   41409225 Pi Sin[----]
             2                    2
> -------------------- + ------------------------ +
        2097152                536870912


            phi0 18                   phi0 20
 147744025 Pi Sin[----]   2133423721 Pi Sin[----]
               2                     2
> ----------------------- + ------------------------- +
        2147483648                34359738368
```

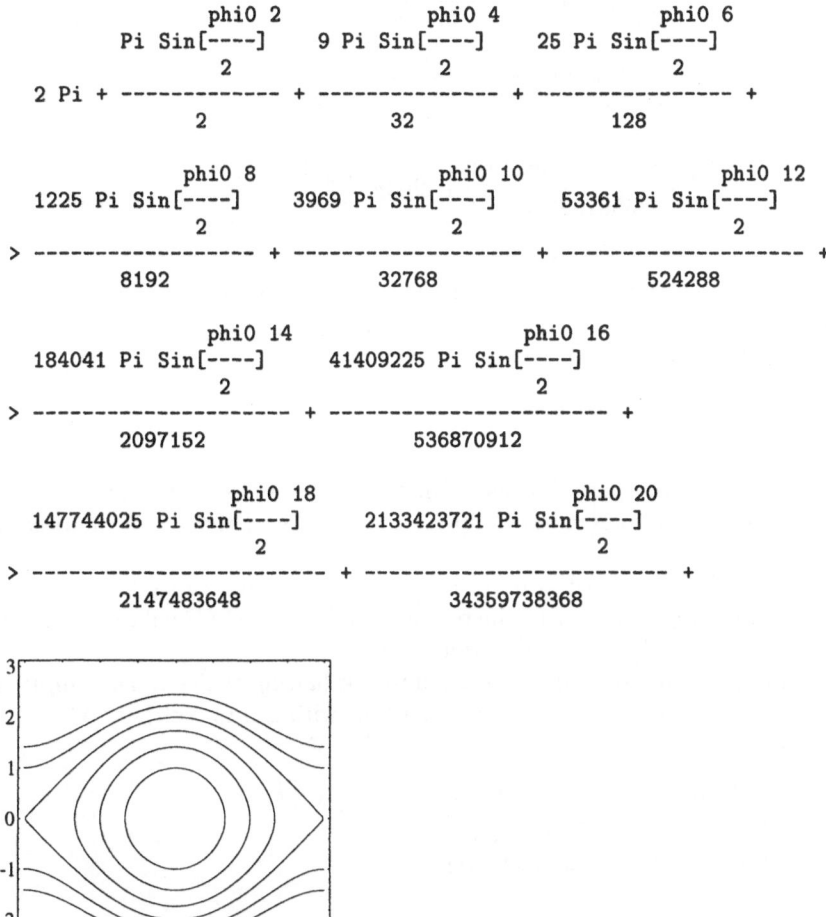

Fig. 1.4. Phase-space plot of $\dot{\varphi}$ versus φ for different energies $E/(mgl) = -1/2$, 0, $1/2$, 1, $3/2$, and 2 (from the inside out)

```
      phi0 2 11
> O[Sin[----] ]
        2
```

Therefore, in the future, we will give the corresponding TEX form for results of this kind. This form can, by the way, be obtained in *Mathematica* with the command TeXForm and can be translated by using TEX or LATEX:

$$2\pi + \frac{\pi \sin^2 \frac{\varphi_0}{2}}{2} + \frac{9\pi \sin^4 \frac{\varphi_0}{2}}{32} + \frac{25\pi \sin^6 \frac{\varphi_0}{2}}{128} + \frac{1225\pi \sin^8 \frac{\varphi_0}{2}}{8192}$$

$$+ \frac{3969\pi \sin^{10} \frac{\varphi_0}{2}}{32768} + \frac{53361\pi \sin^{12} \frac{\varphi_0}{2}}{524288} + \frac{184041\pi \sin^{14} \frac{\varphi_0}{2}}{2097152}$$

$$+ \frac{41409225\pi \sin^{16} \frac{\varphi_0}{2}}{536870912} + \frac{147744025\pi \sin^{18} \frac{\varphi_0}{2}}{2147483648}$$

$$+ \frac{2133423721\pi \sin^{20} \frac{\varphi_0}{2}}{34359738368} + \mathcal{O}(\sin^2 \frac{\varphi_0}{2})^{11} .$$

A comparison with the exact function $T(\varphi_0)$ reveals that the error of this approximation is of the order of at most a few percent as long as φ_0 is less than 120°; beyond that, it increases significantly.

Exercises

1. Calculate the period T as a function of φ_0, once using EllipticK and then with NIntegrate. Compare the processing times and the precision of the results.
2. What is the contribution of the higher harmonics to $\varphi(t)$ as a function of the amplitude φ_0 ($\hat{=}$ increasing nonlinearity)? Calculate the Fourier coefficients $|b_s|/|b_2|$ as functions of φ_0.
3. Program the following algorithm for evaluating the complete elliptic integral $K(\sin \alpha)$ and compare the result with EllipticK[$\sin^2 \alpha$]:

   Start: $a_0 = 1$, $b_0 = \cos \alpha$,
   Iteration: $a_{i+1} = (a_i + b_i)/2$, $b_{i+1} = \sqrt{a_i b_i}$,
   Stop: $|a_n - b_n| < \varepsilon$,
   Verify: EllipticK[$\sin^2 \alpha$] $\simeq \dfrac{\pi}{2a_n}$.

Literature

Baumann G. (1996) Mathematica in Theoretical Physics: Selected Examples from Classical Mechanics to Fractals. TELOS, Santa Clara, CA

Crandall R.E. (1991) Mathematica for the Sciences. Addison-Wesley, Redwood City, CA

Zimmerman R.L., Olness F.I., Wolpert D. (1995) Mathematica for Physics. Addison-Wesley, Reading, MA

1.3 Fourier Transformations

Surprisingly often, physics problems can be described to a good approximation by linear equations. In such cases, the researcher has a powerful tool at his or her disposal: the decomposition of the signal into a sum of harmonic oscillations. This tool, which has been thoroughly investigated mathematically, is the *Fourier transformation*. The transformation can be formulated in a particularly compact way with the help of complex-valued functions. Almost any signal, even a discontinuous one, can be represented as the limit of a sum of continuous oscillations. An important application of Fourier transformations is the solution of linear differential equations. The expansion of a function in terms of simple oscillations plays a big role not only in physics, but also in image processing, signal transmission, electronics, and many other areas.

Frequently, data are only available at discrete points in time or space. In this case, the numerical algorithms for the Fourier transformation are particularly fast. Because of this advantage, we want to investigate the Fourier transformation of discrete data here. In the following sections we will use it to smooth data, to calculate electrical circuits, and to analyze lattice vibrations.

Mathematics

Let a_r, $r = 1, \ldots, N$ be a sequence of complex numbers. Its Fourier transform is the sequence b_s, $s = 1, \ldots, N$, with

$$b_s = \frac{1}{\sqrt{N}} \sum_{r=1}^{N} a_r \, \exp \left[2\pi \mathrm{i} \frac{(r-1)(s-1)}{N} \right] . \qquad (1.8)$$

This formula has the advantage of being almost symmetric in a_r and b_s, since the inverse transformation is

$$a_r = \frac{1}{\sqrt{N}} \sum_{s=1}^{N} b_s \, \exp \left[-2\pi \mathrm{i} \frac{(r-1)(s-1)}{N} \right] . \qquad (1.9)$$

Thus, the signal $\{a_1, \ldots, a_N\}$ has been decomposed into a sum of oscillations

$$c_s(r) = \frac{b_s}{\sqrt{N}} \, \exp \left[-\mathrm{i}\omega_s (r-1) \right] \qquad (1.10)$$

with frequencies $\omega_s = 2\pi(s-1)/N$. The coefficient $b_s = |b_s| \exp(\mathrm{i}\varphi_s)$ consists of an amplitude and a phase. Both (1.8) and (1.9) can be extended to all integers r and s. Then because $\exp(2\pi ik) = 1$ for $k \in \mathbb{Z}$, a_r and b_s are periodic in r and s respectively, with a period N:

$$a_r = a_{r+kN} , \quad b_s = b_{s+kN} , \tag{1.11}$$

so in each sum the index r or s can run through any interval of length N. The smallest oscillation frequency is $\omega_2 = 2\pi/N$ ($\cong$ "wavelength" N); all other frequencies are integer multiples of ω_2. The largest frequency is $\omega_N = 2\pi(N-1)/N$, as $b_{N+1} = b_1$ and therefore $c_{N+1}(r)$ has the same functional values as $c_1(r)$.

If all signals a_r are real, their Fourier coefficients b_s have the following symmetry:

$$b_s^* = \frac{1}{\sqrt{N}} \sum_{r=1}^{N} a_r \exp\left[-2\pi i \frac{(r-1)(s-1)}{N}\right]$$

$$= \frac{1}{\sqrt{N}} \sum_{r=1}^{N} a_r \exp\left[2\pi i \frac{(r-1)(-s+2-1)}{N}\right] = b_{-s+2} .$$

Here, b^* is the complex conjugate of b. From this we can conclude that

$$b_s = b_{-s+2}^* = b_{N-s+2}^* ; \tag{1.12}$$

thus, only the values $b_1, \ldots, b_{N/2+1}$ are relevant, and $b_{N/2+2}, \ldots, b_N$ are redundant.

A useful property of Fourier transforms becomes evident in the convolution of two sequences $\{f_r\}$ and $\{g_r\}$. In this operation, the elements f_r are weighted with the g_j and then added up. One can interpret this as local averaging and use it, for example, to mathematically treat the effect of measurement errors. The convolution is defined as follows:

$$h_r = \sum_{j=1}^{N} f_{r+1-j}\, g_j . \tag{1.13}$$

Let the sequences $\{\tilde{h}_s\}$, $\{\tilde{g}_s\}$, and $\{\tilde{f}_s\}$ be the corresponding Fourier transforms. Combining the Fourier expansions

$$\tilde{h}_s = \frac{1}{\sqrt{N}} \sum_{r=1}^{N} h_r \exp\left[2\pi i \frac{(r-1)(s-1)}{N}\right] ,$$

$$g_j = \frac{1}{\sqrt{N}} \sum_{m=1}^{N} \tilde{g}_m \exp\left[-2\pi i \frac{(m-1)(j-1)}{N}\right] ,$$

$$f_{r+1-j} = \frac{1}{\sqrt{N}} \sum_{n=1}^{N} \tilde{f}_n \exp\left[-2\pi i \frac{(n-1)(r-j)}{N}\right]$$

according to (1.13) yields

$$\tilde{h}_s = \left(\frac{1}{\sqrt{N}}\right)^3 \sum_{r,j,m,n} \tilde{g}_m \, \tilde{f}_n$$

$$\times \exp\left[2\pi i \frac{(r-1)(s-1) - (m-1)(j-1) - (n-1)(r-j)}{N}\right] .$$

The sums over r and j can be evaluated

$$\sum_{l=1}^{N} \exp\left(2\pi i \frac{lq}{N}\right) = N\,\delta_{q,kN} , \quad k \in \mathbb{Z} ,$$

and we are left with

$$\tilde{h}_s = \sqrt{N} \sum_{m,n} \tilde{g}_m \tilde{f}_n \exp\left(2\pi i \frac{m-s}{N}\right) \delta_{s-n,0}\,\delta_{n-m,0} = \sqrt{N}\tilde{g}_s\tilde{f}_s . \quad (1.14)$$

Thus, after the transformation, the convolution turns into a simple product. The inverse transformation then yields the convoluted function

$$h_r = \sum_{s=1}^{N} \tilde{g}_s \, \tilde{f}_s \, \exp\left[-2\pi i \frac{(r-1)(s-1)}{N}\right] . \quad (1.15)$$

In the next section, we will use the convolution to smooth experimental data.

Algorithm

In *Mathematica*, one obtains a list of the Fourier coefficients {b1, ..., bN} via the command

    Fourier[{a1, ..., aN}]

and their inverse by using

    InverseFourier[{b1, ..., bN}]

It may be interesting anyway to look into the *Fast Fourier Transform* (FFT) algorithm. The most straightforward way of calculating all series elements b_s according to (1.8) amounts to multiplying a matrix of the form

$$W_{s,r}(N) = \frac{1}{\sqrt{N}} \exp\left[2\pi i \frac{(r-1)(s-1)}{N}\right]$$

by a vector $\boldsymbol{a} = (a_1, \ldots, a_N)$; this takes N^2 steps. The FFT algorithm, however, can accomplish the same in a number of steps that increases only as $N \log N$. To this end, the sum (1.8) is split up into two partial sums with odd and even indices r:

$$b_s(N) = \sum_{t=1}^{N/2} W_{s,2t-1}(N)\, a_{2t-1} + \sum_{t=1}^{N/2} W_{s,2t}(N)\, a_{2t}$$

$$= \sum_{t=1}^{N/2} \frac{1}{\sqrt{2}} W_{s,t}\left(\frac{N}{2}\right) a_{2t-1}$$

$$+ \exp\left(2\pi i \frac{s-1}{N}\right) \sum_{t=1}^{N/2} \frac{1}{\sqrt{2}} W_{s,t}\left(\frac{N}{2}\right) a_{2t}$$

$$= \frac{1}{\sqrt{2}}\left[b_s^o\left(\frac{N}{2}\right) + \exp\left(2\pi i \frac{s-1}{N}\right) b_s^e\left(\frac{N}{2}\right)\right]. \tag{1.16}$$

Here, $b_s^o(N/2)$ and $b_s^e(N/2)$ are the Fourier transforms of the coefficients $\{a_1, a_3, \ldots, a_{N-1}\}$ and $\{a_2, a_4, \ldots, a_N\}$, i.e., of two sequences with $N/2$ elements each (the superscripts o and e stand for odd and even). By repeating the split for each of the two parts, one obtains b_s^{oo}, b_s^{oe}, b_s^{eo}, and b_s^{ee}. This process is repeated until one is left with only one a_r value which corresponds to some b_s, e.g., $b_s^{eoeeoeoo}$; this last assignment can be specified. With some additional bookkeeping tricks, this allows us to program a fast algorithm.

Because this algorithm only works optimally if N is a power of 2, we assume in the following that N is of the form $N = 2^j$. To determine $F(N)$, the number of steps required to calculate the Fourier transform of an N-element sequence, one can derive the following recursive relation from (1.16):

$$F(N) = 2\,F\left(\frac{N}{2}\right) + k\,N. \tag{1.17}$$

The term $2\,F(N/2)$ is immediately obvious, as it corresponds to the calculation of $b_s^o(N/2)$ and $b_s^e(N/2)$. The index s in (1.16) goes from 1 to N, where it has to be noted that the values of $b_s^o(N/2)$ and $b_s^e(N/2)$ are cycled through twice because these sequences have a period $N/2$. A term proportional to N is added because N phase factors $\exp[2\pi i(s-1)/N]$ have to be calculated and N multiplications and additions have to be performed. This term can also be used to account for the sorting of the $\{a_r\}$ into odd and even terms. It is a simple exercise to prove that $F(N) = N\left[F(1) + k \log_2 N\right]$ solves the recursive relation (1.17); thus, since the initial value $F(1) = 0$, $F(N) = k\,N \log_2 N$.

Application

As an example, we want to investigate the Fourier transformation of a rectangular pulse of width 2τ,

$$f(t) = \frac{1}{2}\left[\operatorname{sign}(\tau + t) + \operatorname{sign}(\tau - t)\right]. \tag{1.18}$$

We begin by defining the function f in the interval $[-T/2, T/2]$, with $T/2 > \tau$, and then continue it periodically with a period T. The Fourier expansion of this function is

$$f(t) = \sum_{n=-\infty}^{\infty} \tilde{f}_n \exp(in\omega t) ,$$ (1.19)

where $\omega = 2\pi/T$ and the $\tilde{f}_n$ are determined by integrals over $f(t)$

$$\tilde{f}_n = \frac{1}{T} \int_{-T/2}^{T/2} f(t) \exp(-in\omega t) \, dt .$$ (1.20)

For a rectangular pulse, this integral is easily executed; the result is

$$\tilde{f}_n = \frac{2}{T} \frac{\sin(n\omega\tau)}{n\omega} .$$ (1.21)

Depending on the period T, the function $\sin(x)/x$ is evaluated at different points $2\pi n\tau/T$, and the coefficients $\tilde{f}_n$ vanish if $2n\tau/T$ is an integer.

Now we probe $f(t)$ for N discrete values $t_r = -T/2 + r\,T/N$ within a period and apply **Fourier** to the sequence of numbers $\{a_r = f(t_r)\}_{r=1}^{N}$ to obtain its Fourier transform $\{b_1, b_2, \ldots, b_N\}$. We want to elucidate the relationship between the exact Fourier coefficients $\tilde{f}_n$ and the b_s from the discrete Fourier transformation. To this end, we replace the functional values $a_r = f(t_r)$ in (1.8) by the infinite Fourier series (1.19) with $\omega = 2\pi/T$. The summation over r yields Kronecker symbols and for N even we obtain the following relation:

$$\frac{b_s}{\sqrt{N}} = \exp\left[i\pi\frac{(N-2)(s-1)}{N}\right] \sum_{k=-\infty}^{\infty} \tilde{f}_{1-s+kN} .$$ (1.22)

A comparison of the absolute values yields

$$\frac{|b_s|}{\sqrt{N}} = \left| \sum_{k=-\infty}^{\infty} \tilde{f}_{1-s+kN} \right| .$$ (1.23)

Since $|\tilde{f}_n| \to 0$ as $n \to \infty$, we have $|b_s|/\sqrt{N} \approx |\tilde{f}_{1-s}|$ for $s = 1, 2, \ldots, N/2$. The question whether $|b_s|/\sqrt{N} < |\tilde{f}_{1-s}|$ or $|b_s|/\sqrt{N} > |\tilde{f}_{1-s}|$, the effect of the so-called *aliasing*, cannot be answered in general. This depends on the phase between $\tilde{f}_{1-s}$ and primarily $\tilde{f}_{1-s\pm N}$. In the example investigated here, one can realize one case as well as the other by varying the parameter τ/T. For $N = 64$ and $\tau/T = 1/10$, Fig. 1.5 shows the absolute value of $\tilde{f}_{1-s}$ as a solid curve and the absolute values of $b_s/\sqrt{N}$ as filled circles. The inverse transformation from the coefficients b_s to the original values $a_r = f(t_r)$ is done according to (1.9), using $r = N(t_r/T + 1/2)$. This equation yields the exact values of the rectangular pulse at the discrete points t_r. On the other hand, (1.9) can be interpreted as an approximation to the entire function $f(t)$, if t_r is replaced by continuous t values. Figure 1.6 shows, in the plot on the left, that this does not yield a good result. The reason for the low quality of the approximation is the strongly oscillating contribution from the high frequencies corresponding to $N/2 < s \le N$. In (1.11) in the mathematics

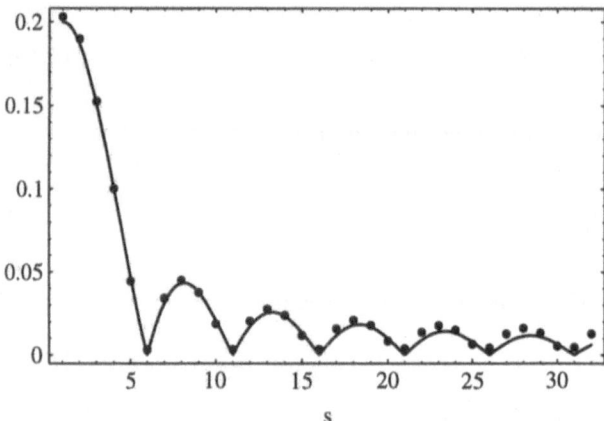

Fig. 1.5. Continuous and discrete Fourier transformations of a rectangular pulse. The solid curve shows $|\tilde{f}_{1-s}|$ as a function of s, the filled circles are the values of $|b_s|N^{-1/2}$

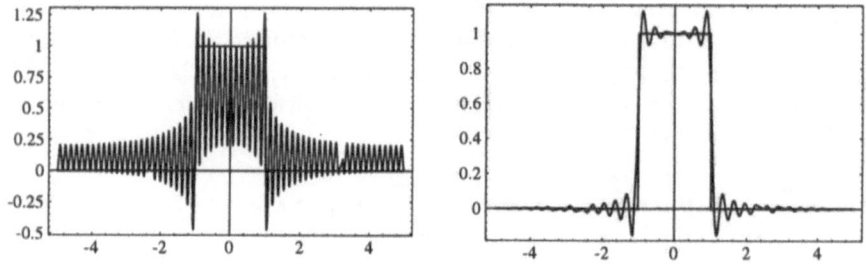

Fig. 1.6. Less-than-optimal (*left*) and improved (*right*) continuation of the discrete inverse Fourier transformation to continuous t values

part above, we have shown that the sum can be taken over any interval of length N. We take advantage of this freedom to choose an interval that is symmetric to the origin and use only the lowest frequencies:

$$f(t_r) = \frac{1}{\sqrt{N}} \sum_{s=-N/2+1}^{N/2} b_s \exp\left[-2\pi \mathrm{i}\frac{(r-1)(s-1)}{N}\right]$$

$$= \frac{1}{\sqrt{N}} \sum_{s=1}^{N/2} b_s \exp\left[-2\pi \mathrm{i}\frac{(r-1)(s-1)}{N}\right]$$

$$+ \frac{1}{\sqrt{N}} \sum_{s'=1}^{N/2} b_{s'-N/2} \exp\left[-2\pi \mathrm{i}\frac{(r-1)(s'-N/2-1)}{N}\right]$$

$$= \frac{1}{\sqrt{N}} \sum_{s=1}^{N/2} b_s \exp\left[-2\pi \mathrm{i}\frac{(r-1)(s-1)}{N}\right]$$

$$+\frac{1}{\sqrt{N}} \sum_{s'=1}^{N/2} b^*_{N/2-s'+2} \exp\left[-2\pi i \frac{(r-1)(s'-N/2-1)}{N}\right].$$

If this version is extended to continuous t values, the step is approximated much better, as shown in the right-hand plot of Fig. 1.6. The discontinuities of the rectangular pulse $f(t)$, however, can only be reproduced at the expense of introducing strong oscillations on both sides.

Exercises

1. **Peak voltage.** A time-dependent voltage is probed at $N = 64$ points in time resulting in the values $U_1, \ldots, U_N$. Assume that its Fourier transform, the sequence $\{b_1, \ldots, b_N\}$, has the power spectrum $|b_s|^2 = |b_{N-s+2}|^2 = 1$ for $s = 9, 10, \ldots, 17$, $b_s = 0$ otherwise. If the voltage signal $\{U_1, \ldots, U_N\}$ is to be real, we have to have $b_{N-s+2} = b_s^*$.
   (a) What does the voltage signal $\{U_1, \ldots, U_N\}$ look like if the phases of the Fourier coefficients are constant, e.g., $b_s = 1$, if $b_s \neq 0$?
   (b) Choose random phases for the non-zero b_s, i.e., $b_s = \exp(i\varphi_s)$ with random φ_s. What does the signal look like now?
   (c) **Competition:** The objective is to find phases that result in the lowest possible peak voltage, i.e., $\min_{\{\varphi_s\}} \max_r |U_r|$.

2. **Aperiodic crystal.** Assume a chain of atoms with a periodic arrangement.

   $$1\,1\,0\,0\,1\,1\,0\,0\,1\,1\,0\,0\,1\,1\,0\,0\,.$$

   By X-ray diffraction one obtains the absolute values of the Fourier coefficients $|b_s|$ of the function a_r that is defined as

   $$a_r = \begin{cases} 1 & \text{if atom 1 is at the position } r, \\ 0 & \text{otherwise}. \end{cases}$$

   (a) Plot the Fourier spectrum b_s.
   (b) Next, generate a random sequence of 1s and 0s and plot its Fourier spectrum.
   (c) Finally, use the following algorithm to generate an aperiodic crystal and compare its Fourier spectrum to those of the other two crystals.

      (i) Start with: 0
                         1
                         1 0
      (ii) Generate the next line by appending the next-to-last line to the last one.
      (iii) Iterate this procedure up to the desired length.

   The number of atoms in the nth line is the Fibonacci number F_n.

Literature

Crandall R.E. (1991) Mathematica for the Sciences. Addison-Wesley, Redwood City, CA

Press W.H., Teukolsky S.A.,Vetterling W.T., Flannery B.P. (1992) Numerical Recipes in C: The Art of Scientific Computing. Cambridge University Press, Cambridge, New York

DeVries P.L. (1994) A First Course in Computational Physics. Wiley, New York

Wolfram S. (1996) The Mathematica Book, 3rd ed. Wolfram Media, Champaign, IL, and Cambridge University Press, Cambridge

1.4 Smoothing of Data

Experimental data are usually affected by a noticeable statistical error. The scatter of the measured values can be eliminated to a certain degree by averaging over adjacent data points. Smoothing data this way is particularly useful when they are presented graphically. Here we want to demonstrate the procedure using data with an artificially generated scatter. To smooth these data, we weight, for each point in turn, its neighbors by their distance and add them up. To avoid generating new structures in the process, a "Gaussian bell curve" is used as the weight function. In other words, the data are convoluted with a Gauss function, a smoothing operation for which the Fourier transformation from the previous section can be used.

Physics

Let $f(t)$ be a physical quantity that is measured at discrete times t_i. The exact values $f_i = f(t_i)$ are modified by random numbers r_i such that the measured values are given by

$$g_i = f_i + r_i \, , \ i = 1, 2, 3, \ldots, N \, . \tag{1.24}$$

For each i we want to calculate new values $\bar{g}_i$ that are mean values over a vicinity of i. Then the sequence $\{\bar{g}_i\}$ will be much smoother and a better representation of the original function $f(t)$ than the sequence of the g_i values.

To calculate $\bar{g}_i$, one could simply add up the g_j in an interval about i. There is a better method, however, that does not generate any unwanted additional structures. In this method, a weighted average is used that takes into account the neighbors according to their distance $|j - i|$ from the point i. Very distant neighbors are only weighted weakly. As a weight function, or *kernel*, a "Gaussian bell curve" is used. Let k_j be such a weight function, which is restricted to $j = 1, \ldots, N$ as before. The k_j values for $j < 1$ are determined by the periodicity relation (1.11). Of course, the kernel has to be normalized:

$$\sum_{j=1}^{N} k_j = 1 \,.\tag{1.25}$$

Then $\bar{g}_r$ is constructed as follows:

$$\bar{g}_r = \sum_{j=1}^{N} g_{r-j+1} k_j \,.\tag{1.26}$$

Thus, $\bar{g}$ is the discrete convolution of the functions g and k that has been discussed in Sect. 1.3. Therefore, it can simply be written as a Fourier transform,

$$\bar{g}_r = \sum_{s=1}^{N} \tilde{g}_s \tilde{k}_s \exp\left[-2\pi i \frac{(s-1)(r-1)}{N}\right]\,,\tag{1.27}$$

where $\tilde{g}$ and $\tilde{k}$ are the Fourier transforms of g and k respectively.

Algorithm

We generate the data we want to smooth as an array data in *Mathematica*, using the values of the Bessel function $J_1(x)$, and add noise in the form of random numbers.

```
data=
Table[N[BesselJ[1,x] + 0.2 (Random[]-1/2)], {x, 0, 10, 10/255}]
```

For the kernel, we pick a Gaussian of width σ:

```
kernel = Table[N[Exp[-x^2/(2*sigma^2)]], {x,-5,5,10/255}]
```

This, however, puts the largest functional values in the middle of the array kernel. Therefore, we have to rotate this array cyclically to shift the maximum to the beginning of the array, i.e., to kernel[[1]]. This is done via the command

```
kernel = RotateLeft[kernel,127]
```

This kernel still has to be normalized according to (1.25):

```
kernel = kernel/Apply[Plus, kernel]
```

Then, using (1.14), one simply obtains the smoothed data (1.27) from the expression

```
Sqrt[256] InverseFourier[Fourier[data] Fourier[kernel]]
```

In this form, the product of two arrays yields the array of the products of the individual elements, just as required by (1.27).

Results

Figure 1.7 shows the results of the calculations done with $\sigma = 0.4$. The smoothed data are compared to the noisy data and to the original function. The curve $\overline{g}_i$ is indeed smooth and has a somewhat smaller amplitude than the original curve. Of course, this is to be expected since the smoothed curve averages over the vicinity and therefore is always lower than the maximum values of the data. Particularly the periodicity of the data that is tacitly assumed here by using the Fourier transformation leads to deviations near the edges.

We would like to mention that this very example is presented in Sect. 3.8.3 of the *Mathematica* book. In the second edition of this handbook, however, that section contains several errors: the kernel is not normalized, the factor $\sqrt{N}$ is omitted from the convolution, and only the right half of the "bell curve" is used for the kernel. In the third edition these mistakes have been corrected. Finally, we want to point out that one has to be careful when doing the averaging. If too small a width is chosen for the kernel, it is obvious that one will obtain the original data without smoothing. Too wide a weight function, on the other hand, not only averages out the scatter but also destroys small-scale structures of the original function. In the extreme case of an infinitely wide kernel, the smoothed function has one single, constant value. Therefore the width of the weight function needs to be adjusted carefully to the data to be smoothed.

If the statistics of the measurement uncertainties are known, there are theoretical models that can be used to calculate the optimum kernel. Alternatively, one can fit the data by as smooth a curve as possible that yields the correct χ^2 value (see next section). For example, one can draw polynomials through adjacent interpolation nodes and require their curvatures to be minimal. We cannot treat this broad area of statistical data analysis here; instead, we have to refer to the appropriate textbooks.

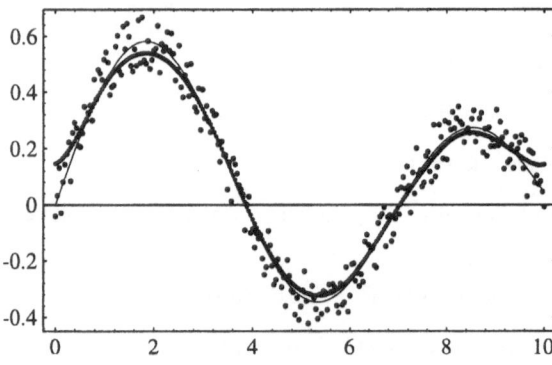

Fig. 1.7. Original Bessel function (*thin curve*) and noisy data (*circles*). The smoothed data points appear as a bold curve

Exercise

The result of the algorithm for smoothing data introduced above depends on the width σ chosen for the kernel. Calculate the smoothed data for a wide range of σ values. If the width of the kernel is less than the distance between the data points, one obtains essentially the original data. If, on the other hand, the width is larger than the entire x-interval, the smoothed data are nearly constant. Determine a value for σ that appears appropriate to you.

Literature

Press W.H., Teukolsky S.A., Vetterling W.T., Flannery B.P. (1992) Numerical Recipes in C: The Art of Scientific Computing. Cambridge University Press, Cambridge, New York

Wolfram S. (1996) The Mathematica Book, 3rd ed. Wolfram Media, Champaign, IL, and Cambridge University Press, Cambridge

1.5 Nonlinear Fit

One goal of science is to find regularities in measured data and to express them mathematically. Therefore, model functions with free parameters are frequently fitted to the data. The models are – hopefully – the results of theories; the computer is needed to find the "best" parameter values and provide a measure of the quality of the fit. To this end, comprehensive tools from the mathematical field of statistics, as well as equally comprehensive program packages, are available to the researcher.

If the models are linear in the parameters, the mathematical theory for this is particularly well developed. Frequently, however, the physicists' models are more complex, requiring nonlinear fits. Here, we want to use a simple example to demonstrate the principle of nonlinear parameter search. In doing so, we will estimate the quality of the fit by a χ^2 test, in which we have to set a measure of our confidence ourselves. We will never be able to prove that our model represents reality; the test will, however, allow us to exclude false models with a high degree of probability. For an accepted model, we will also be able to specify a measure for the precision of the parameters.

Using the example of a damped oscillation that is measured at just eleven points in time, yielding very noisy data, we fit the data by a model with four parameters and determine their errors.

Theory

Assume N data pairs $\{Y_i, t_i\}$, to which a model function $g(a, t_i)$ with an unknown M-component parameter vector a is to be fitted. Assume further that it is known that for each t_i the data Y_i are measured with errors ε_i that vanish when averaged over many experiments and have the variances σ_i^2, i.e., $\langle \varepsilon_i \rangle = 0$ and $\langle \varepsilon_i^2 \rangle = \sigma_i^2$. The "best" set of parameters a is defined to be the vector a_0 whose components minimize the quadratic deviation χ^2, where χ^2 is defined as follows:

$$\chi^2(a) = \sum_{i=1}^{N} \left[\frac{Y_i - g(a, t_i)}{\sigma_i} \right]^2 . \tag{1.28}$$

The distribution of $\chi^2(a_0)$ is known if $g(a_0, t_i)$ is a suitable model and the errors $\varepsilon_i = Y_i - g(a_0, t_i)$ are uncorrelated and follow a Gaussian distribution. One can verify by repeating the experiment many times that the probability for χ_0^2 is given by

$$P_{N-M}\left(\chi_0^2\right) = \frac{1}{\Gamma\left(\frac{N-M}{2}\right)} \int_0^{\chi_0^2/2} e^{-t} t^{\frac{N-M}{2}-1} dt . \tag{1.29}$$

The integral is known in mathematical literature as an incomplete gamma function $\gamma((N-M)/2, \chi_0^2/2)$. $\chi^2(a_0)$ only has $N-M$ independent variables, since M degrees of freedom are fixed by the minimum condition. For large values of $N-M$, the central limit theorem applies: χ^2 is distributed according to a Gaussian with mean $\langle \chi^2 \rangle = N - M$ and variance $2(N-M)$.

Now what is the meaning of the χ^2 distribution (1.29) for the precision of our fit? Let us assume that our experiment yields the value $\chi_0^2 = \chi^2(a_0)$ for the minimum of (1.28), with a value $P(\chi_0^2) = 0.99$. This means that if the model is correct the value of χ^2 would be less than χ_0^2 in 99% of all experiments. We could still accept that as a valid fit, since we cannot exclude the possibility that our experiment belongs to those 1% of all cases for which χ^2 is greater than or equal to χ_0^2. If, however, we had $P = 0.9999$, our fit would only be correct if it belonged to the 0.01% of the experiments for which $\chi^2 \geq \chi_0^2$. In this case we would certainly conclude that our assumptions, and our model in particular, are not correct. On the other hand, the value of χ_0^2 must not be too small, since our data points do have statistical errors. The probability for an experiment to yield a smaller value than our χ_0^2 is again given by the distribution (1.29). The interval for the value of $P(\chi_0^2)$ that we want to accept is called the *confidence interval*. Where we set the limit, at 1% or at 0.01%, depends on ourselves and on our experience with similar problems.

Let us assume now that we are satisfied with the quality of our fit, i.e., $\chi^2(a)$ has a minimum for some parameters a_0 and the value of $\chi^2(a_0)$ falls within our confidence interval. If we could repeat the experiment several times, we would of course obtain different errors ε_i and consequently a different parameter vector a_0. We perform just one experiment, however, but want to use the data to estimate the error of a_0 anyway.

Here, the χ^2 distribution (1.29) helps us again, as contours in M-dimensional a-space with constant values of $\chi^2(a)$ are a measure of the error of a_0. For small deviations $|a - a_0|$, these contours are ellipsoids with M principal axes. Their lengths are a measure of the change one can allow for a_0 along these axes before the fit is no longer acceptable.

One can even quantify this statement. To this end, we generate artificial data Y_i by adding errors ε_i that are distributed according to a Gaussian with $\langle \varepsilon_i^2 \rangle = \sigma_i^2$ to the model $g(a_0, t_i)$: $Y_i = g(a_0, t_i) + \varepsilon_i$. With these data, we run the fit procedure again, i.e., we search for a new minimum a_1, using the simulated data. If repeated several times, this yields the parameter vectors $a_1, a_2, \ldots$. From the width of the distribution of each component of the a_k we obtain the error bars for the fit parameters a_0.

If the deviation $|a - a_0|$ is small enough to permit truncating the expansion of $\chi^2(a)$ (from the experimental data) about a_0 after the quadratic term, it can be shown that the quantity $\Delta = \chi^2(a) - \chi^2(a_0)$ is again distributed according to the distribution function P from (1.29), this time with M rather than $N - M$ degrees of freedom. Therefore, if we require as before that our fit belong to those 99% of all possible fits closest to the correct model, then the inequality

$$P_M \left[\chi^2(a) - \chi^2(a_0) \right] \leq 0.99 \tag{1.30}$$

determines the region of allowed values of a. In parameter space, the regions of constant Δ are ellipsoids. The projection of this $(M - 1)$-dimensional surface onto the axis i then yields the error interval for the parameter a_i.

Algorithm

For the nonlinear fit, too, it is easiest to use the functions available in *Mathematica*. To find the minimum of $\chi^2(a)$, we use the function NonlinearRegress from the package Statistics`NonlinearFit`. It offers various possibilities for entering data and initial conditions; in addition, one can change the method of the minimum search and have the program display intermediate results. Of course, one can also provide one's own definition of $\chi^2(a)$ and use FindMinimum to find the value a_0.

To generate the data, and for our model, we choose a damped sinusoidal oscillation

```
f[t_] := a Sin[om t + phi] Exp[-b t]
```

with four parameters $a = $ {a, om, phi, b}. This oscillation is measured at 11 points in time t_i for the parameter set $a = \{1, 1, 0, 0.1\}$, and noise in the form of uniformly distributed random numbers is added to the data:

```
data =
Table[{t, Sin[t]Exp[-t/10.] + 0.4*Random[] - 0.2}//N,
                              {t, 0, 3Pi, 0.3Pi}]
```

For the errors ε_i (which are not distributed according to a Gaussian) we have

$$\langle \varepsilon_i \rangle = 0, \quad \sigma_i^2 = \langle \varepsilon_i^2 \rangle = \frac{1}{0.4} \int\limits_{-0.2}^{0.2} x^2 \, dx = \frac{2}{150} \cdot \simeq .013 \qquad \sigma_i = \sqrt{.013}$$

The search for a minimum is facilitated if we can provide an approximate set of values for a_0. After loading the statistics packages needed via

```
Needs["Statistics`Master`"]
```

we call the minimum search routine:

```
NonlinearRegress[data,f[t],t,
              {{a,1.1},{om,1.1},{phi,.1},{b,.2}},
              ShowProgress->True]
```

The χ^2 distribution is available in *Mathematica* as well, in the package `Statistics`ContinuousDistributions`. It has the self-explanatory name `ChiSquareDistribution[...]`. For the argument, the number of degrees of freedom has to be provided, which in our case, with $N = 11$ data points and $M = 4$ parameters is given by $N - M = 7$. By using PDF one obtains the density distribution, i.e., the integrand of $P_7(\chi_0^2)$, (1.29), whereas CDF yields the integral, i.e., $P_7(\chi_0^2)$ itself. With `Quantile`, the function $P_7(\chi_0^2)$ is inverted. Thus,

```
Quantile[ChiSquareDistribution[7], 0.95]
```

yields the value of χ_0^2 that is defined by the condition that 95% of all experiments will result in a χ^2 value smaller than χ_0^2. If two parameters are kept fixed, one can use `ContourPlot` to display, for the remaining two parameters, the area in parameter space for which $\chi^2(a) = \chi^2(a_0) + \Delta$.

Results

Figure 1.8 shows the function `f[t]` for a=1, om=1, phi=0, and b=0.1, as well as the 11 data points that were obtained from `f[t]` with the random errors ε_i. Starting from a=1.1, om=1.1, phi=0.1, and b=0.2, NonlinearRegress finds the minimum a_0 of $\chi^2(a)$ in about eight steps. In order to extract the best fit parameters from the result, we have to enter the command `BestFitParameters /. NonlinearRegress[...]`, yielding

$$\{a \to 0.824, \ om \to 0.976, \ phi \to 0.024, \ b \to 0.068\} \qquad (1.31)$$

The χ^2 distribution is shown in Fig. 1.9. Strictly speaking, the distribution (1.29) is valid only for Gaussian-distributed errors ε_i, but we do not expect a large difference for our uniformly distributed errors ε_i. For the correct set of parameters, this density function yields the distribution of χ_0^2 for different experiments, i.e., for different realizations of the ε_i.

We obtain the confidence interval for the value of χ_0^2 from

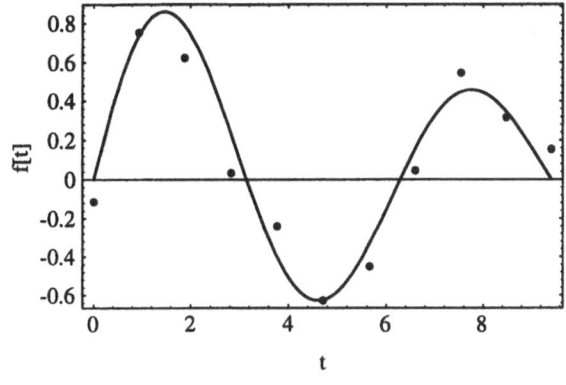

Fig. 1.8. Damped oscillation and noisy data

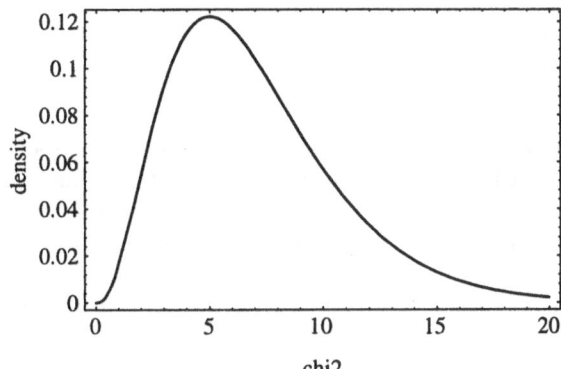

Fig. 1.9. χ^2 density distribution with seven degrees of freedom

```
limit[x_] = Quantile[ChiSquareDistribution[7],x]
```

and the result of {limit[0.05], limit[0.95]} is {2.2, 14.1}. This means that, given a large number of experiments (with the correct parameters), 5% of them should yield a value $\chi_0^2 \geq 14.1$ and 5% a value $\chi_0^2 \leq 2.2$. In our case, we obtain $\chi_0^2 = 9.4$, well inside the confidence interval; therefore, we have no reason to doubt the result of our fit.

Contours for $P_4(\chi^2(a) - \chi_0^2) = 0.68$ and $P_4(\chi^2(a) - \chi_0^2) = 0.90$ can be seen in Fig. 1.10. These are three-dimensional surfaces in four-dimensional parameter space, which is why we can only plot slices. Of course, the optimal parameter set is situated exactly in the center of all these contour surfaces. The (a,b) slice shows that variations of the amplitude a can be compensated for by a change in b^{-1}, the time constant of the damping, without reducing the quality of the fit. Therefore, one cannot simply specify the cut along the optimum value of b for the error in a, but has to use the projection onto the a-axis instead. In particular, the left-hand panel of Fig. 1.10 shows that it is not possible to determine an upper limit for the time constant b^{-1} since the contour of the outer confidence region passes through b=0 (i.e., $b^{-1} = \infty$).

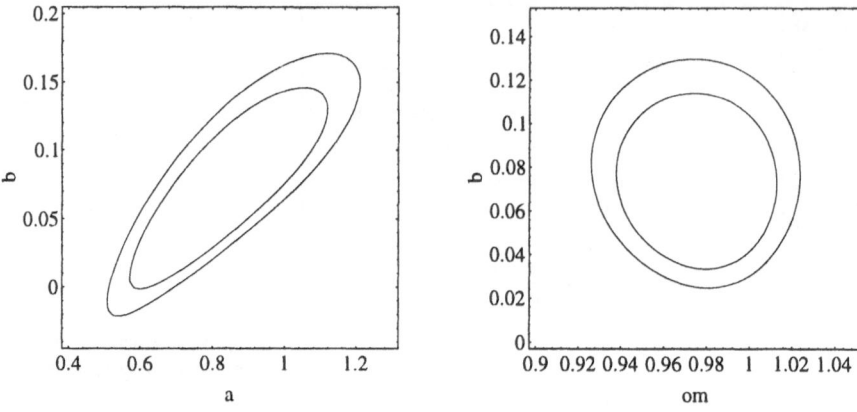

Fig. 1.10. Contours of constant $\chi^2(a)$. Intersection with the **a**-**b** plane in parameter space (*left*) and with the om-**b** plane (*right*)

If one varies the frequency om rather than the amplitude **a** the compensation discussed above does not happen, as shown in the right-hand panel of Fig. 1.10. The true values, om=1 and b=0.1, lie close to the edge of the inner confidence region; therefore, the confidence interval should not be too narrow. If one plots f[t] for two extreme values from Fig. 1.10, namely a=1.3, b=0.15 and a=0.5, b=0, one can see in Fig. 1.11 that both curves still represent the data relatively well. The small number of data ($N = 11$) and the large error do not permit a better fit.

As discussed in the theory section, there is an alternative method for estimating the errors of the parameters. One takes the optimum parameter set a_0 for one experiment and uses it to generate new, artificial data that are fitted in turn. Repeating this for the same a_0 yields a set of parameter vectors a_k. Figure 1.12 shows the results of 100 iterations, using the a_0 from (1.31), together with the contours from Fig. 1.10. It can be seen that in the (a,b) projection even more than 90% of the data fall within the outer contour

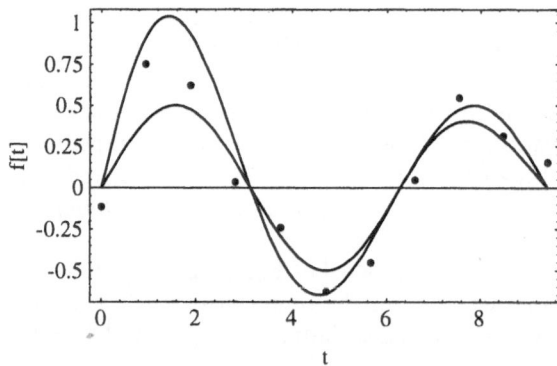

Fig. 1.11. Two fit curves generated from parameters from the outer contour of the **a**-**b** plot shown in Fig. 1.10

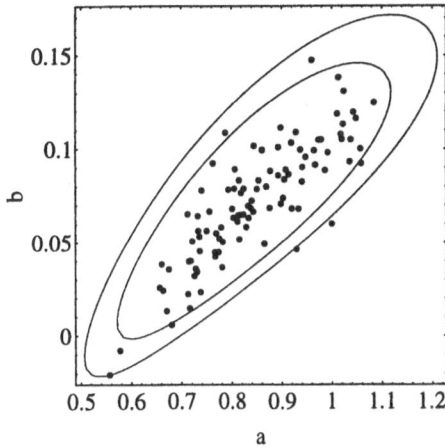

Fig. 1.12. Parameter values a and b from the fits of the artificial data are shown together with the contours from the left-hand plot of Fig. 1.10

of the slice. In order to specify error limits for a_0, we calculate $\chi^2(a_k)$ for each of the 100 parameter vectors a_k and the original data set $\{Y_i\}$ and sort the a_k by increasing χ^2. For a confidence level of 68%, for example, we then take into account only the first 68 of these vectors, the last of which sets a χ^2 limit $\chi^2_{68} = \chi^2(a_{68})$. All these remaining a_k then lie inside the quasi-ellipsoid $\chi^2(a) \leq \chi^2_{68}$. We project the points a_k onto the axes of parameter space to obtain the extreme dimensions of the ellipsoid. In this way, we can eventually make the statement that with 68% certainty the true vector a_{true} lies in a quasi-ellipsoid that is contained in the rectangular parallelepiped

$$a = 0.82 \pm 0.26, \quad b = 0.07 \pm 0.06, \quad om = 0.98 \pm 0.05, \quad phi = 0 \pm 0.2 .$$

Of course, not every point a of the rectangular parallelepiped belongs to the 68% region, owing to the correlations mentioned above. Whether or not it belongs to this region can be determined from its $\chi^2(a)$.

Exercise

The data $\{x_i, f(x_i)\}$ in the file **twinpeak.dat** are the result of the function

$$f(x_i) = \exp\left(-\frac{x_i^2}{2}\right) + a \exp\left[-\frac{(x_i - b)^2}{2\sigma^2}\right] + r_i ,$$

where the r_i are normally distributed random numbers with mean 0 and width $\sigma_0 = 0.05$. You can load the data during a *Mathematica* session via

```
data = << twinpeak.dat
```

and display them with **ListPlot[data]**. Find the amplitude a, the position b, and the width σ of the additional peak and specify the errors of the three fit parameters.

Literature

Bevington P.R., Robinson D.K. (1992) Data Reduction and Error Analysis for the Physical Sciences. McGraw-Hill, New York

Honerkamp J. (1994) Stochastic Dynamical Systems: Concepts, Numerical Methods, Data Analysis. VCH, Weinheim, New York

Press W.H., Teukolsky S.A., Vetterling W.T., Flannery B.P. (1992) Numerical Recipes in C: The Art of Scientific Computing. Cambridge University Press, Cambridge, New York

Wolfram S. (1996) The Mathematica Book, 3rd ed. Wolfram Media, Champaign, IL, and Cambridge University Press, Cambridge

1.6 Multipole Expansion

The expansion of physics equations in terms of a small quantity is an important tool of the theoretical physicist. In this process, expressions may result that are complicated and nonintuitive. Frequently, one has no feel for the expansion's deviation from the true value.

The multipole expansion of an electrostatic potential, a favorite chapter from the course on electrodynamics, is a simple example of this. From far away, a charge distribution looks like a point charge. As the observer gets closer, the dipole moment will become noticeable. On even closer approach the quadrupole moment will be noticeable as well. This approximate description can be put into a compact mathematical form using a scalar, a vector, and a tensor. Now – with *Mathematica* – we can program this just as compactly, but also display it graphically and investigate the deviation from the exact potential.

Physics

The electrostatic potential $\Phi(r)$ of N pointlike charges e_i at the positions $r^{(i)}$ is given by

$$\Phi(r) = \sum_{i=1}^{N} \frac{e_i}{|r - r^{(i)}|} \ . \tag{1.32}$$

The electric field is determined by the gradient of Φ:

$$E = -\nabla\Phi(r) \ , \nabla\Phi = \left(\frac{\partial\Phi}{\partial x}, \frac{\partial\Phi}{\partial y}, \frac{\partial\Phi}{\partial z} \right) \ . \tag{1.33}$$

When observing this potential from a long distance, i.e., for $|r - r^{(i)}| \to \infty$, one can expand $\Phi(r)$:

$$\Phi(r) = \frac{q}{r} + \frac{p \cdot r}{r^3} + \frac{1}{2r^5} r Q r + \mathcal{O}\left(\frac{1}{r^4}\right) \ , \tag{1.34}$$

where $r = |\boldsymbol{r}|$. Here, q is the total charge, $\boldsymbol{p}$ the dipole moment, and $\mathbf{Q}$ the quadrupole tensor:

$$q = \sum_{i=1}^{N} e_i \ , \quad \boldsymbol{p} = \sum_{i=1}^{N} e_i \boldsymbol{r}^{(i)} \ , \quad Q_{kl} = \sum_{i=1}^{N} e_i \left[3 r_k^{(i)} r_l^{(i)} - \delta_{kl} r^{(i)2} \right] \ , \quad (1.35)$$

where $r_k^{(i)}$ is the kth component of $\boldsymbol{r}^{(i)}$.

Algorithm

We want to investigate this expansion using an example of five positive and five negative unit charges that we place in the x–y plane, randomly distributed within a square centered at the origin. Thus, we generate ten vectors by

```
rpoint:={2 Random[]-1, 2 Random[]-1, 0}
Do[r[i] = rpoint, {i,10}]
```

Each vector r[i] is an array of three numbers. The first five points shall have positive charges, the last five negative ones. Now we want to graphically display these charges. To do this, we first use the function Line[...] to draw plus and minus signs at the corresponding places, eliminating the z-coordinate by using Drop[r[i],-1]:

```
p1 = Graphics[Table[Line[{Drop[r[i],-1]-{0.08,0},
                Drop[r[i],-1]+{0.08,0}}],{i,5}]]
p2 = Graphics[Table[Line[{Drop[r[i],-1]+{0,0.08},
                Drop[r[i],-1]-{0,0.08}}],{i,5}]]
p3 = Graphics[Table[Line[{Drop[r[i+5],-1]-{0.08,0},
                Drop[r[i+5],-1]+{0.08,0}}],{i,5}]]
```

Next, we use Circle[ ] to draw a circle around each symbol,

```
p4 = Graphics[{Thickness[0.001],
          Table[Circle[Drop[r[i], -1], 0.1],{i,10}]}]
```

and finally we use Show[p1,p2,p3,p4,*options*] to plot all four graphics objects together. Figure 1.13 shows the result.

To be able to specify the potential, we first define the distance between two vectors,

```
dist[r_,s_] = Sqrt[(r-s).(r-s)]
```

The dot between two arrays of numbers – *Mathematica*'s extended notation for this is Dot[11,12] – effects a scalar product of the two vectors, i.e., the sum of the products of corresponding components of the vectors. Without the dot, on the other hand, the vectors would be multiplied element by element, resulting in an array. Now, one can directly enter $\Phi(\boldsymbol{r})$ from (1.32), using $e_i = \pm 1$ and:

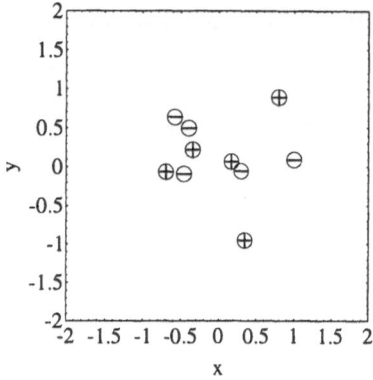

Fig. 1.13. Positive and negative charges randomly distributed in the x–y plane

```
pot[rh_] := Sum[1/dist[rh,r[i]]-1/dist[rh,r[i+5]],{i,5}]
```

We present three ways to visualize this result: first we use `Plot3D` to plot the potential in the form of valleys and mountains rising out of the x–y plane; next, we plot the contours of this mountain range by using `ContourPlot`; finally, we calculate the electric field and plot it via `PlotVectorField`, from the package `Graphics`PlotField``. In order to see anything in the latter case, however, one has to normalize the field, i.e., display only its direction.

The dipole and quadrupole moments can be formulated just as easily in *Mathematica*. According to (1.35), we have

```
dipole = Sum[r[i] - r[i+5],{i,5}]
quadrupole[r_] :=
    Table[3 r[[k]] r[[l]] - If[k==1,r.r,0],{k,3},{l,3}]
qsum = Sum[quadrupole[r[i]]-quadrupole[r[i+5]],{i,5}]
```

Here, the Kronecker symbol δ_{kl} is represented by the following expression:

```
If[k==1,1,0]
```

To calculate the magnitude of a vector r, which we could also express as `dist[r,0]`, we use the function

```
magn[r_]=Sqrt[r.r]
```

With this, the expansion (1.34) of the potential $\Phi(r)$ can be defined immediately. In our example, there is no net charge ($q = 0$); therefore, the first term `pot1` is the dipole term, and `pot2` contains the quadrupole term in addition:

```
pot1[r_] = dipole.r / magn[r]^3
pot2[r_] = pot1[r]+1/2/magn[r]^5 r.qsum.r
```

Here, `qsum` is an array of arrays, in this case a 3×3 matrix; r is an array of numbers, i.e., a vector. The function `Dot[ , ]`, indicated by a dot (.) for brevity, calculates contractions of tensors with an arbitrary number of indices (= nested arrays). Therefore, in this case the quadratic form "vector times matrix times vector" can be written very easily as r.qsum.r. In a conventional programming language like C, by contrast, one would have to nest two for loops:

```
sum = 0;
for(i=0; i<3; i++) {
    for(j=0; j<3; j++) {
        sum = sum + r[i]*qsum[i][j]*r[j] }}
```

One obtains the electric field E according to (1.33) as follows:

```
efield =
-{D[pot[{x,y,z}],x],D[pot[{x,y,z}],y],D[pot[{x,y,z}],z]}
```

To obtain its direction, one has to divide by $|E|$:

```
direction = efield/magn[efield]
```

Results

In our simple example, five positive and five negative charges $|e_i| = 1$ are randomly distributed in the x–y plane (Fig. 1.13). In this plane they generate the potential pot[{x,y,0}] that is displayed as a "mountain range" in Fig. 1.14 and by its contours ($\Phi = $ const.) in Fig. 1.15.

We now want to compare the dipole and quadrupole approximations to the exact potential. To this end, we look at a path in the x–y plane ($z = 0$) parallel to the y-axis at $x = 0.6$. This path passes close to two positive charges at $y \simeq \pm 1$ and a negative charge and a dipole near $y \simeq 0$. Figure 1.16 shows

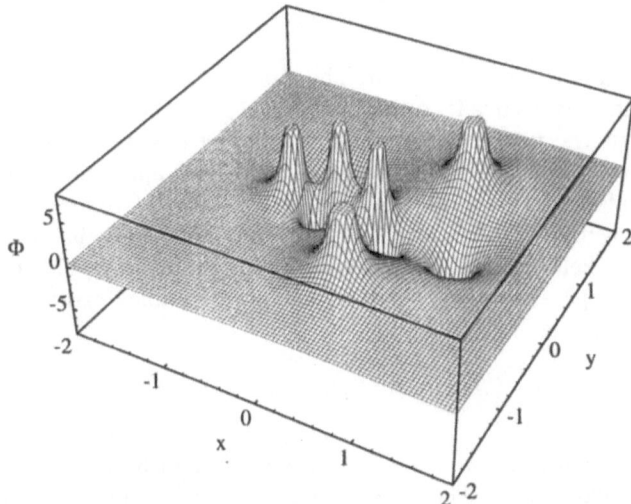

Fig. 1.14. The potential Φ of the ten unit charges in the plane $z = 0$

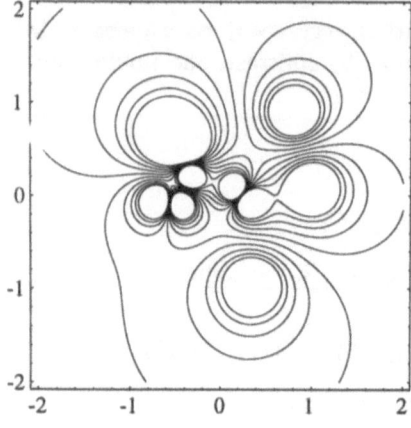

Fig. 1.15. The same potential as in Fig. 1.14, but displayed as a contour plot

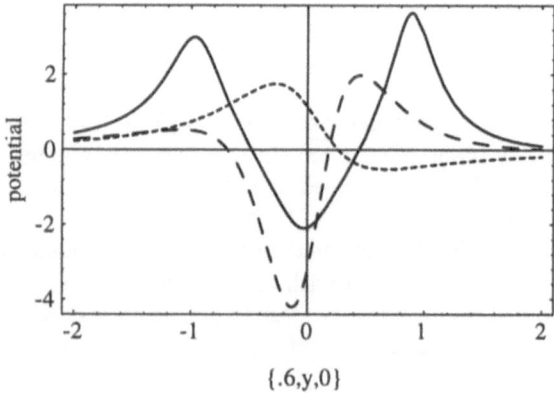

{.6,y,0}

Fig. 1.16. The exact potential (*solid curve*), the dipole term (*short-dashed curve*), and the quadrupole approximation (*long-dashed curve*)

the result. The solid curve is the exact result. The dipole term (short-dashed) can change its sign only once since it is proportional to $p \cdot r$; therefore it can not correctly represent the two positive maxima near the positive charges. Even with the quadrupole correction (long-dashed curve) the approximation describes the potential near the charges only in a rough, qualitative manner. The direction of the electric field can be seen in Fig. 1.17. At a long distance from the charges, the field rotates around the origin, as for a dipole. In the vicinity of the charges, on the other hand, a more complex structure is observed.

Exercise

The magnetic field generated by a current I flowing through a conducting circular loop is to be calculated. The conductor is assumed to be situated in the x–y plane, with its center at the origin, and to have a radius a. The vector potential A, from which the magnetic field $B = \nabla \times A$ is calculated,

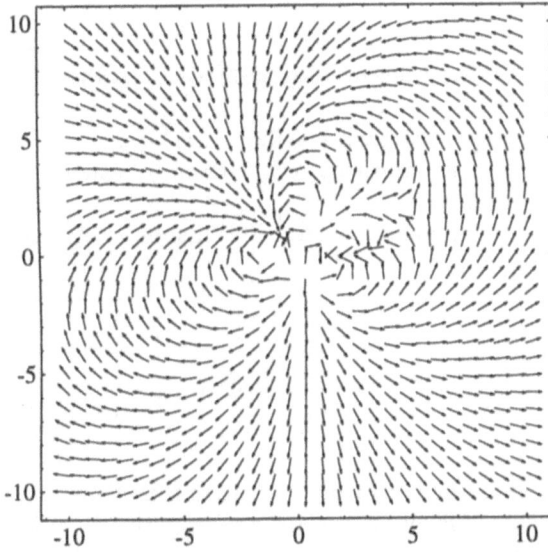

Fig. 1.17. The direction of the electric field of the ten unit charges in the plane $z = 0$

takes a particularly simple form in spherical coordinates (r, θ, ϕ), with unit vectors e_r, e_θ, e_ϕ. Only its ϕ component is non-zero, and for this component the textbook by J.D. Jackson gives the expression

$$A_\phi(r, \theta) = \frac{\mu_0}{\pi} \frac{I\,a}{\sqrt{a^2 + r^2 + 2ar\sin\theta}} \left[\frac{(2 - k^2)\,K(k) - 2E(k)}{k^2} \right]$$

with

$$k^2 = \frac{4ar\sin\theta}{a^2 + r^2 + 2ar\sin\theta} \; .$$

$K(k) = $ `EllipticK[k^2]` and $E(k) = $ `EllipticE[k^2]` are the complete elliptic integrals of the first and second kind.

Calculate the magnetic field $B(x, y, z)$ and plot its direction in the x–z plane, using the function `PlotVectorField[...]`. Attempt to plot the field lines of B in the x–z plane by integrating a suitable differential equation.

Hint (applies only to *Mathematica* versions before 3.0): The problem that *Mathematica* does not know how to calculate the derivatives `EllipticK'` and `EllipticE'` can be solved since the derivatives of the elliptic integrals can be expressed again by elliptic integrals. The two lines

```
EllipticK'[x_]=1/(2x)*(EllipticE[x]/(1-x)-EllipticK[x])
EllipticE'[x_]=1/(2x)*(EllipticE[x]-EllipticK[x])
```

in your program will solve the problem.

Literature

Gradshteyn I.S., Ryzhik I.M. (1994) Table of Integrals, Series, and Products. Academic Press, Boston, MA

Jackson J.D. (1975) Classical Electrodynamics. Wiley, New York

Smith C., Blachman N. (1995) The Mathematica Graphics Guidebook. Addison-Wesley, Reading, MA

Wickham-Jones T. (1994) Mathematica Graphics: Techniques & Applications. TELOS, Santa Clara, CA

1.7 Line Integrals

Physics

Work = force × path length. This seemingly simple equation from high school physics immediately turns more difficult if one realizes that the force is a vector field $\boldsymbol{F}(\boldsymbol{r})$, i.e., it assigns a vector to each point in space $\boldsymbol{r}$, and that the path P is a curve $\boldsymbol{r}(t)$ in space that can be parametrized by the time t. Thus, the work W is given by the line integral

$$W = \int_P \boldsymbol{F} \cdot \mathrm{d}\boldsymbol{r} = \int_{t_i}^{t_f} \boldsymbol{F}\left(\boldsymbol{r}\left(t\right)\right) \cdot \frac{\mathrm{d}\boldsymbol{r}}{\mathrm{d}t}\mathrm{d}t \ . \tag{1.36}$$

This means that the scalar product force · velocity is integrated over the time interval $[t_i, t_f]$. This integral becomes simple, if $\boldsymbol{F}(\boldsymbol{r})$ is the gradient of a potential $\Phi(\boldsymbol{r})$

$$\boldsymbol{F} = -\boldsymbol{\nabla}\Phi \ . \tag{1.37}$$

In this case, $W = \Phi\left[r(t_i)\right] - \Phi\left[r(t_f)\right]$. In general, however, the evaluation of the line integral can be cumbersome.

   Mathematica offers the possibility to perform vector analysis conveniently even in symbolic form. We want to demonstrate this for a line integral; for additional calculations and vector operations in other coordinate systems, we refer to the package `Calculus`VectorAnalysis``.

Algorithm and Result

First, we define three paths in space $\boldsymbol{r}(t)$, all of which lead from $\{1, 0, 0\}$ to $\{1, 0, 1\}$:

```
r1 = {Cos[2Pi t], Sin[2Pi t], t}
r2 = {1, 0, t}
r3 = {1 - Sin[Pi t]/2, 0, (1 - Cos[Pi t])/2}
```

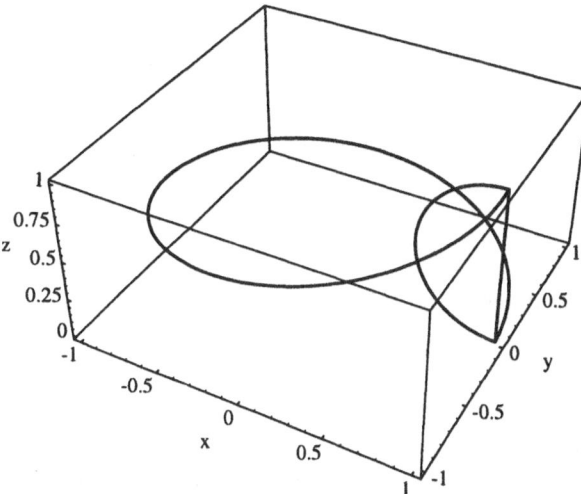

Fig. 1.18. Three different integration paths

The first path is a spiral, the second the straight line connecting the two points, and the third a semicircle in the x–z plane. `ParametricPlot3D` displays these paths, as shown in Fig. 1.18. Then we define a vector field, namely

```
f[{x_,y_,z_}]={2x y + z^3, x^2, 3x z^2}
```

The velocity at which the path is traversed can simply be defined by the derivative operator `D[...]`; this operator is `Listable`, i.e., it acts on list elements:

```
v[r_]:=D[r,t]
```

The line integral can even be defined as a function that acts on paths:

```
int[r_]:=Integrate[k[r].v[r],{t,0,1}]
```

Now we enter `int[r1]`, `int[r2]`, and `int[r3]` and obtain the same result, the value 1, for all three paths. The independence of the integral from the integration path makes us suspect that F is the gradient of a potential. In that case, according to the laws of vector analysis, the curl of F, which is defined as

```
curl[{fx_, fy_, fz_}]:= { D[fz,y] - D[fy,z],
                          D[fx,z] - D[fz,x],
                          D[fy,x] - D[fx,y] }
```

has to vanish. Indeed, `curl[f[{x,y,z}]]` yields the null vector `{0,0,0}`.

How does one find the potential $\Phi(r)$, which is uniquely determined up to a constant? According to (1.36)

$$\Phi(r) = -\int_0^r \boldsymbol{F} \cdot d\boldsymbol{s} = -\int_0^1 \boldsymbol{F} \cdot \frac{d\boldsymbol{s}}{dt} dt \,, \tag{1.38}$$

where we choose s(t) = r4 = t {x,y,z} as a path from 0 to r. Except for the sign, int[r4] yields the potential $\Phi(x, y, z) = -(x^2 y + x z^3)$.

We can now slightly modify the field $\boldsymbol{F}(r)$, for example to

f[{x,y,z}] = {2x y^2 + z^3, x^3, 3x z^3}

In this case we obtain a different result for each path: after we apply Simplify, int[r1] yields

$$1 + \frac{3}{4\pi^2} + \frac{3\pi}{4} \,,$$

int[r2] yields 3/4, and for int[r3] we obtain the value $3/4 + 9\pi/256$. If we now calculate the curl again, then curl[f[{x,y,z}]] yields a value different from **0**, in agreement with the dependence of the integral on the integration path.

Exercises

1. Calculate the lengths of the curves r1, r2, and r3.
2. Choose the parameterization $t = \tau^2$ for the paths r1, r2, r3, calculate the three line integrals and plot the absolute values of the three accelerations $\left|d^2 r/dt^2\right|$ as functions of τ.
3. Test the equality of the line integrals for a different vector field $a(x, y, z)$, with $a = \nabla\Phi$ and Φ a potential of your choice.

Literature

Dennery P., Krzywicki A. (1996) Mathematics for Physicists. Dover Publications, Mineola, NY

1.8 Maxwell Construction

Frequently in physics the mutual dependence of quantities is governed by nonlinear equations that can only be solved numerically. One example of this is the $p(V, T)$ curve of a van der Waals gas, with the temperature T as a parameter, which obtains physical significance through the so-called *Maxwell construction*.

At low temperatures, the isotherms of the equation of state, which is known as the van der Waals equation, have S-shaped bends, the *van der Waals loops*, that are prohibited for reasons of thermodynamics. They have to be replaced by horizontal lines that in a sense bisect the loops. In order to construct these so-called two-phase lines, one has to numerically solve a nonlinear equation that, furthermore, contains an integral. In this way, one obtains a description of the phase transition from gas to liquid.

Physics

For an ideal gas of N noninteracting classical particles with no internal degrees of freedom, the theory of heat provides a simple relationship between volume V, pressure p, and temperature T:

$$pV = Nk_{\mathrm{B}}T \ . \tag{1.39}$$

Here, k_{B} is the Boltzmann constant. Taking into account the interaction between the particles, one obtains, to a simple approximation, the van der Waals equation

$$\left(p + \frac{a}{V^2}\right)(V - b) = Nk_{\mathrm{B}}T \tag{1.40}$$

with parameters a and b.

This equation specifies, for example, the isotherms, i.e., the pressure p as a function of the volume V for constant temperatures T. For high temperatures, the pressure decreases as the volume increases, whereas for low temperatures $p(V)$ increases again over a certain range (see Fig. 1.19). There is, thus, a critical temperature T_{c} below which this van der Waals loop appears. For $T < T_{\mathrm{c}}$, a phase transition between gas and liquid takes place; at the precise temperature T_{c} the difference between the liquid and gaseous phases vanishes.

T_{c} and V_{c} are determined by the condition that the first and second derivatives of $p(V,T)$ with respect to V vanish,

$$\frac{\partial p}{\partial V}(V,T) = 0 \ \text{ and } \ \frac{\partial^2 p}{\partial V^2}(V,T) = 0 \ , \tag{1.41}$$

with the result

$$T_{\mathrm{c}} = \frac{8a}{27Nk_{\mathrm{B}}b} \ , \quad V_{\mathrm{c}} = 3b \ , \quad p_{\mathrm{c}} = \frac{a}{27b^2} \ . \tag{1.42}$$

Now, if one normalizes p, V, and T with $p_{\mathrm{c}}, V_{\mathrm{c}}$, and T_{c} respectively, one obtains an equation that contains no more parameters ($\tilde{x} = x/x_{\mathrm{c}}$):

$$\left(\tilde{p} + \frac{3}{\tilde{V}^2}\right)\left(3\tilde{V} - 1\right) = 8\tilde{T} \ . \tag{1.43}$$

Thus, for $\tilde{T} < 1$, this equation yields the unphysical van der Waals loop, i.e., over a certain range three volumes $\tilde{V}_1, \tilde{V}_2$, and $\tilde{V}_3$ correspond to each pressure $\tilde{p}$. For thermodynamic reasons, the transition from the gas, with a large volume $\tilde{V}_3$, to the liquid, with a small volume $\tilde{V}_1$, takes place at the pressure $\tilde{p}_{\mathrm{t}}$ for which

$$\int_{\tilde{V}_1}^{\tilde{V}_3} \tilde{p}(\tilde{V}) d\tilde{V} = \tilde{p}_t \left(\tilde{V}_3 - \tilde{V}_1 \right) . \tag{1.44}$$

At $\tilde{p}_t$, gas and liquid are simultaneously present for all volumes $\tilde{V}$ with $\tilde{V}_1 \le \tilde{V} \le \tilde{V}_3$; one observes a two-phase mixture. Geometrically, the equation above means that the area between the curve $\tilde{p}(\tilde{V})$ and the straight line $\tilde{p}_t = $ constant in the range $\tilde{V}_1$ to $\tilde{V}_2$ is the same as the corresponding area between $\tilde{V}_2$ and $\tilde{V}_3$ (see Fig. 1.19). Using *Mathematica*, we want to construct the curve $\tilde{p}_t = $ constant, the two-phase line.

Algorithm and Result

First, we verify (1.42) for the critical point. We enter the van der Waals equation (1.40) $(N k_B T = t)$:

```
p=t/(v-b) - a/v^2
```

and define the two equations (1.41):

```
eq1=D[p,v]==0
eq2=D[p,{v,2}]==0
```

Here, the symbol = indicates an assignment, whereas == tests the equality and is executed first. We solve these equations for t and v:

```
sol=Solve[{eq1,eq2},{t,v}]
```

which results in a rule:

```
         8 a
{{ t -> ----, v -> 3b}}
        27 b
```

Since the solution is not always unique, Solve yields a list of rules; hence the double curly braces. We obtain the critical pressure by applying the inner part of the rule to p,

```
pc = p/.sol[[1]]
```

with the result a/(27 b^2).

Now we define the normalized function $\tilde{p}(\tilde{V})$ from (1.43):

```
p[v_] = 8t/(3v-1)-3/v^2
```

For the two-phase line we need two equations in order to determine the two unknown volumes v1 and v3 between which gas and liquid coexist. The first equation says that coexistence requires equality of pressures:

```
eq3=p[v1]==p[v3]
```

The second one equates the two areas according to (1.44):

```
eq4=p[v1]*(v3-v1)==Integrate[p[v],{v,v1,v3}]
```

An attempt with Solve indicates that *Mathematica* does not find an analytic solution. Therefore, we will use FindRoot to obtain a numerical solution. To find a reasonable result, however, the initial values of the (v1,v3) search have to be specified rather precisely. Therefore, for temperatures T < 1 the function plot[T_] first determines those values of v where p[v] attains its local minimum and maximum respectively and calculates the arithmetic mean of the two, vtest. Then the *Mathematica* command Solve[p[v]==p[vtest],v] returns three solutions; the two outer ones are suitable initial values. The final Maxwell function then is

```
pmax[v_]:= If[v < v1 || v > v3, p[v], p[v1]]
```

Thus, if v is between the solutions v1 and v3, the two-phase line is returned, otherwise it is p[v]. Figure 1.19 shows the result for various values of the normalized temperature $\tilde{T}$, together with the unphysical functions p[v]. As v → 1/3, the pressure diverges. The top curve describes $p(V)$ in the gas phase, and the next lower one represents the situation just at the critical temperature. The three lowest curves show the separation into a liquid (small volume) and a gaseous phase. Along the two-phase line, liquid and gas coexist in thermal equilibrium.

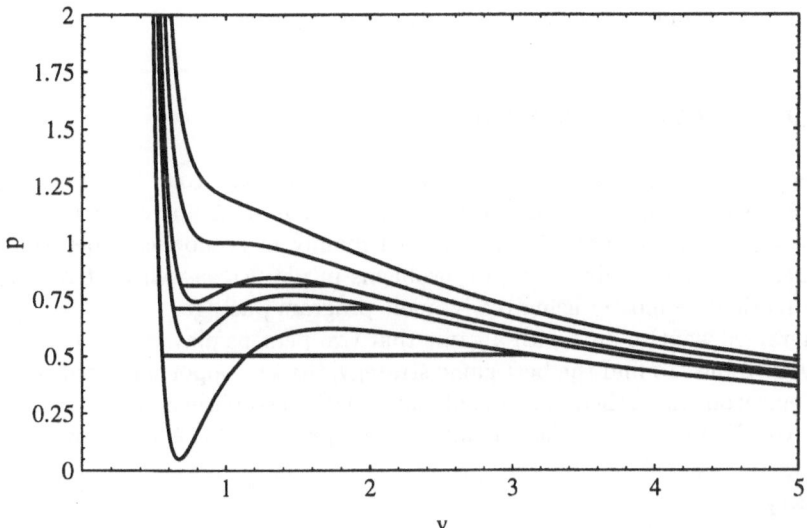

Fig. 1.19. Maxwell construction for the van der Waals equation. Isotherms for the normalized temperatures $\tilde{T} = 1.05, 1.0, 0.95, 0.92, 0.85$ (*top to bottom*)

Exercise

Consider a quantum particle of mass m in a one-dimensional potential well

$$V(x) = \begin{cases} -V_0 & \text{for } -a \leq x \leq a, \\ 0 & \text{otherwise}. \end{cases}$$

From the quantum mechanics course you know that the energy levels E of the bound states are given by the following equations:

$$\varepsilon = \frac{E}{V_0} = \frac{q^2}{x^2} - 1,$$

where $x = (a/\hbar)\sqrt{2mV_0}$, and q is a solution of the transcendental equation

$$\tan q = \frac{\sqrt{x^2 - q^2}}{q} \quad \text{or} \quad -\cot q = \frac{\sqrt{x^2 - q^2}}{q}.$$

Calculate and plot all energy levels ε_n as a function of the parameter x.

Literature

Boltzmann L. (1898) Vorlesungen über Gastheorie. Barth, Leipzig [English transl. (1995) Lectures on Gas Theory. Dover, New York]

Reif F. (1965) Fundamentals of Statistical and Thermal Physics. McGraw-Hill, New York

Schmid E.W., Spitz G., Lösch W. (1990) Theoretical Physics on the Personal Computer. Springer, Berlin, Heidelberg, New York

1.9 The Best Game Strategy

As our last example of the use of predefined functions from *Mathematica*, we want to look at an optimization problem that does not come directly from physics. Linear optimization is used mainly in economics (operations research), yet repeatedly there are problems in physics as well that require these methods – mostly using ready-made program packages.

Here, we want to deal with a game that two persons play with the aid of a payout table. To find the best game strategy, the two opponents need some theorems from game theory, a field of mathematics established in the twenties by J. von Neumann, and the *Mathematica* program LinearProgramming.

Mathematics

Two people play against each other. After each move, one player has to pay the other an amount determined from a table. Each player attempts to select his moves in such a way that the balance of his gains and losses after many moves will be positive.

For the game, a payout table $K = (K_{ij})$ is needed, e.g.,

$$K = \begin{pmatrix} 0 & 1 & 3 & 1 \\ -1 & 10 & 4 & 2 \\ 7 & -2 & 3 & 7 \end{pmatrix} .$$

Player R can select one of the three rows $i = 1, 2, 3$, while player C has four choices for the columns $j = 1, \ldots, 4$. For a move, player R picks a row number i and player C picks a column number j, while neither knows of the other's choice. Then the numbers are revealed and according to the pair (i, j) player R receives the amount K_{ij} from player C or pays $|K_{ij}|$ if $K_{ij} < 0$. Since the gain of R is the loss of C and vice versa, this game is called the *two-person zero-sum game*. This procedure is repeated many times. At the end, R has received on average the amount K per move. Therefore, R looks for a strategy that yields the largest possible gain K.

C, on the other hand, of course wants to minimize his losses, so he looks for the minimum of K. Thus, for each choice i made by R, he looks for $\min_j K_{ij}$. As he does not know which row R will choose, he might come up with the idea to choose the column that yields the smallest loss in the worst case. For each choice j, C would lose at worst the amount $\max_i K_{ij}$. Therefore, he chooses that value of j which corresponds to $\min_j \max_i K_{ij}$, in this case column $j = 3$. Using similar reasoning, R would choose the row associated with the value $\max_i \min_j K_{ij}$, i.e., $i = 1$. With this strategy, the amount $K_{13} = 3$ would be paid to player R after each move. The limits each player aims for have different values:

$$0 = \max_i \min_j K_{ij} \neq \min_j \max_i K_{ij} = 4 . \tag{1.45}$$

Now what does a gain of 3 mean to player R, who, with this strategy, may expect a minimum "gain" of 0? And should player C, who would have had to count on a loss of 4 in the worst case, be satisfied with 3? Indeed, player R has every reason to rethink his strategy, as closer inspection of the payout table reveals that, if he randomly chooses rows 2 and 3 with the same frequency, his gain will be at least 3 for any choice made by C. And if he slightly prefers row 3 in the process, he will even exceed the gain of 3. One can see from this consideration that it is useful to choose the rows and columns with certain probabilities p_i and q_j respectively. Thus, a strategy for player R consists of the three frequencies $p_1, p_2,$ and p_3 (with $p_1 + p_2 + p_3 = 1$), according to which he chooses the rows without any correlation. On average, he receives the amount

$$K = \sum_{i,j} p_i q_j K_{ij} . \tag{1.46}$$

For any strategy $p_1, p_2,$ and p_3 that R chooses, he will receive the amount $\min_{q_1 \ldots q_4} \sum_{i,j} p_i K_{ij} q_j$ in the worst case. Therefore, R will choose p_1^0, p_2^0, p_3^0 that maximizes the worst-case gain, i.e., he looks for

$$\max_{p_1 \cdots p_3} \ \min_{q_1 \cdots q_4} \ \sum_{i,j} p_i K_{ij} q_j \ . \tag{1.47}$$

Correspondingly, C looks for a strategy $q_1^0 \ldots q_4^0$ that yields the value

$$\min_{q_1 \cdots q_4} \ \max_{p_1 \cdots p_3} \ \sum_{i,j} p_i K_{ij} q_j \ . \tag{1.48}$$

Contrary to the deterministic game (1.45), the optimal stochastic strategies (1.47) and (1.48) yield the same value

$$\max_{p} \min_{q} \sum_{ij} p_i K_{ij} q_j = \min_{q} \max_{p} \sum_{ij} p_i K_{ij} q_j = \sum_{ij} p_i^0 K_{ij} q_j^0 = K_0 \ . \tag{1.49}$$

This is the famous Minimax theorem, which J. von Neumann proved in 1926, at the age of 23. We want to put the meaning of (1.49) in words: if player R chooses an optimal strategy $p_1^0, \ldots, p_3^0$, then his gain after very many moves is at least K_0 for any strategy $q_1, \ldots, q_4$ of player C. This means that for any choice $q_1, \ldots, q_4$ with $q_i \geq 0$ and $q_1 + q_2 + q_3 + q_4 = 1$ we have

$$\sum_{i,j} p_i^0 K_{ij} q_j \geq K_0 \ . \tag{1.50}$$

From this inequality, one can derive a system of four conditions, since with $q_1 = 1$ and $q_2 = q_3 = q_4 = 0$ this becomes $\sum_i p_i^0 K_{i1} \geq K_0$, and correspondingly

$$\sum_i p_i^0 K_{ij} \geq K_0 \quad \text{for all } j \ . \tag{1.51}$$

If we assume that $K_0 > 0$, then with $x_i^0 = p_i^0 / K_0$ this becomes

$$\sum_i x_i^0 K_{ij} \geq 1 \quad \text{for all } j \ . \tag{1.52}$$

Now, according to theorems of linear optimization, the optimal strategy p_1^0, p_2^0, p_3^0 is determined by the maximum value K_0 allowed by the system of inequalities (1.51) or, because $\sum_i x_i^0 = 1/K_0$, by the minimum of $\sum_i x_i^0$.

This fact can be formulated in an even more compact way using vector notation: let K denote the matrix (K_{ij}) and K^T the transpose of K, $\boldsymbol{x} = (x_1, x_2, x_3)^\mathsf{T}$ with $x_i \geq 0$, $\boldsymbol{c} = (1, 1, 1)^\mathsf{T}$, $\boldsymbol{b} = (1, 1, 1, 1)^\mathsf{T}$. Then the optimal strategy $\boldsymbol{p}_0 = K_0 \boldsymbol{x}_0$ and the associated average gain K_0 are determined by a minimum of $\boldsymbol{c} \cdot \boldsymbol{x}$ with the added condition $\mathsf{K}^\mathsf{T} \boldsymbol{x} \geq \boldsymbol{b}$, where the latter vector inequality is meant to be taken component by component. K_0 is given by $K_0 = 1/(\boldsymbol{c} \cdot \boldsymbol{x}_0)$.

According to the duality theorem of linear optimization, K_0 can also be obtained by solving the following problem: Find the maximum of $\boldsymbol{b} \cdot \boldsymbol{y}$ with the added condition $\mathsf{K} \boldsymbol{y} \leq \boldsymbol{c}$ for the vector $\boldsymbol{y} = (y_1, \ldots, y_4)^\mathsf{T}$, with $y_j \geq 0$.

But the solution $\boldsymbol{y}_0$ of this problem yields the optimum strategy for player C, since with $\boldsymbol{q}_0 = K_0 \boldsymbol{y}_0$ one obtains

$$\sum_j K_{ij} q_j^0 \le K_0 \quad \text{for all } i .$$ (1.53)

Then, analogously to (1.50) and (1.51), for any strategy p of R:

$$\sum_{i,j} p_i K_{ij} q_j^0 \le K_0 .$$ (1.54)

This means that in the worst case player C, using an optimal strategy, will lose the amount K_0. If C does not use the strategy q_0, however, his losses can be higher.

One useful theorem should still be mentioned: If one adds the constant d to each matrix element, then p_0 and q_0 remain optimal strategies with the average gain $K_0 + d$. The statements above in reference to (1.51) are strictly valid only for matrices with positive values K_{ij}. Owing to the translation theorem, however, any matrix K can be shifted to positive values; after that, the optimization program is used.

Algorithm and Result

The minimum of a linear function $c \cdot x$ with the added conditions $\mathsf{K}^\mathsf{T} x \ge b$ and $x \ge 0$ can easily be determined using *Mathematica*. For our example we have

```
c = {1, 1, 1}
b = {1, 1, 1, 1}
k = {{0, 1, 3, 1},{-1, 10, 4, 2},{7, -2, 3, 7}}
```

and

```
LinearProgramming[c,Transpose[k],b]
```

yields an optimal vector x_0, from which p_0 is determined by $p_0 = x_0/(c \cdot x_0)$ and K_0 by $K_0 = 1/(c \cdot x_0)$.

Instead of searching for the maximum of $b \cdot y$ we can search for the minimum of $-b \cdot y$ just as well. The inequality $\mathsf{K}y \le c$ is equivalent to $-\mathsf{K}y \ge -c$. Therefore we solve the dual problem by

```
LinearProgramming[-b,-k,-c]
```

This yields the vector y_0 and consequently $K_0 = 1/(b \cdot y_0)$ and $q_0 = K_0 y_0$.
For our example, the calculation gives the results

$$p_0 = (0, 0.45, 0.55)^\mathsf{T} , \quad q_0 = (0.6, 0.4, 0, 0)^\mathsf{T} , \quad K_0 = 3.4 .$$

This means that R uses an optimal strategy if he chooses rows 2 and 3 with the frequencies 0.45 and 0.55 respectively, while C will in no case lose more than $K_0 = 3.4$ if he specifies columns 1 and 2 with the probabilities 0.6 and 0.4 respectively. Both have to make sure, though, that they really make their choice randomly; otherwise, their opponent could react to the correlations and use them to his advantage.

In our example, the algorithm works for negative K_{ij} as well. For other matrices, however, we only found a solution after shifting the matrix K to positive values.

Finally, for entertainment we have written the C program `game.c`, with which the readers of our book can try their luck in a two-person zero-sum game against the computer. In the beginning, the computer uses random numbers to generate a 4×4 payout table with $K_0 = 0$ and, using the program `simplx` from *Numerical Recipes*, calculates the optimal strategies for this matrix. For each move, the player chooses one of the four rows; at the same time, the computer selects one of the columns with the probabilities q_j^0. According to the theory above, the reader could then play without losing on average by an optimal random choice of the rows. Anyone will immediately notice, however, that the computer wins in the long run, unless its opponent calculates the best strategy p_i^0 and acts accordingly. At the end of the game (key `e`) the computer shows how the player would have had to play to avoid losing on average.

Exercise

We want to extend the well-known game of chance with the three symbols rock, paper, scissors to the four symbols rock, paper, scissors, and well. In this version, two players each pick one of these four options and win or lose one point according to the following rules:

- Well swallows rock, well swallows scissors.
- Paper covers well, paper wraps rock.
- Scissors cut paper.
- Rock smashes scissors.

Formulate a payout table for this game and use it to calculate the optimal strategy for both players.

Literature

von Neumann J. (1928) Zur Theorie der Gesellschaftsspiele. Ann. Math. 100:295 [English transl.: Bargmann S. (1959) On the Theory of Games of Strategy. In: Luce R.D., Tucker A.W. (eds.) Contributions to the Theory of Games IV. Princeton University Press, Princeton, NJ, 13–42]

Press W.H., Teukolsky S.A., Vetterling W.T., Flannery B.P. (1992) Numerical Recipes in C: The Art of Scientific Computing. Cambridge University Press, Cambridge, New York

Panik M.J. (1996) Linear Programming: Mathematics, Theory and Algorithms. Kluwer Academic, Dordrecht, New York

2. Linear Equations

Many phenomena in physics can be described by linear equations. In these cases, twice the cause results in twice the effect; this fact allows for a mathematical solution to such problems. Linear systems can frequently be described by vectors, which have sometimes a few, and sometimes a large number of components. In the equations of motion, matrices then appear, whose eigenvalues and eigenvectors describe the energies and the stationary states of the system. Every other form of motion is a superposition of these eigenstates.

There is a multitude of numerical methods of solving linear equations. These highly developed standard algorithms are explained in detail in textbooks on numerical mathematics. In this chapter we have chosen the examples (with the exception of the Hofstadter butterfly) in such a way that they can be examined using predefined functions. When we deal with the electric circuit, the objective is to solve a system of linear equations, and in addition we can use the Fourier transformation again. Other problems, which originate in mechanics as well as quantum physics, lead us to eigenvalue equations, i.e., to the problem of determining eigenvalues and eigenvectors of potentially large matrices.

2.1 The Quantum Oscillator

The equation of motion of quantum mechanics, the Schrödinger equation, is linear: Every superposition of solutions is itself a solution. Therefore, the method of separating the equation into a space- and a time-dependent part by a product ansatz, and later superposing these product solutions is successful. The spatial portion is the so-called stationary Schrödinger equation – an eigenvalue equation which, in the coordinate representation, takes the form of a linear differential equation. The solutions of this equation are wave functions $\Psi(r)$, which assign a complex number to every point r. More specifically, they describe those states of the physical system for which the probability of presence $|\Psi(r)|^2$ does not change with time. To obtain a numerical solution of the Schrödinger equation, one can either approximately discretize the linear differential equation and put it into matrix form, or expand $\Psi(r)$ in terms of a complete set of wave functions $\varphi_n(r)$ and consider only a finite number of them. In both cases, the stationary Schrödinger equation leads to an eigenvalue equation of a finite matrix. We want to investigate the second approach, using the example of the anharmonic oscillator.

Physics

We are investigating a one-dimensional problem, namely a particle of mass m in the quadratic potential $V(q) = m\omega^2 q^2/2$. Here, q is the spatial coordinate of the particle. In dimensionless form, the Hamiltonian of the system is

$$H_0 = \frac{1}{2} \left(p^2 + q^2 \right) . \tag{2.1}$$

Here, the energies are measured in units of $\hbar\omega$, momenta in units of $\sqrt{\hbar m\omega}$, and lengths in units of $\sqrt{\hbar/(m\omega)}$. The eigenstates $|j\rangle$ of H_0 can be found in any quantum mechanics textbook. In the coordinate representation, they are

$$\varphi_j(q) = \left(2^j j! \sqrt{\pi} \right)^{-1/2} e^{-q^2/2} \mathcal{H}_j(q) ,$$

where $\mathcal{H}_j(q)$ are the Hermite polynomials. We have

$$H_0 |j\rangle = \varepsilon_j^0 |j\rangle , \quad \text{where} \quad \varepsilon_j^0 = j + \frac{1}{2} \quad \text{and} \quad j = 0, 1, 2, \ldots . \tag{2.2}$$

The ε_j^0 are the energies of the eigenstates of H_0, and the matrix $\langle j|H_0|k\rangle$ is diagonal because the eigenvalues are not degenerate.

We now add an anharmonic potential to H_0,

$$H = H_0 + \lambda q^4 , \tag{2.3}$$

and try to determine the matrix $\langle j|H|k\rangle$. To do this, it is useful to write q as a sum of creation and annihilation operators $a^\dagger$ and a,

$$q = \frac{1}{\sqrt{2}} \left(a^\dagger + a \right) , \tag{2.4}$$

where a and $a^\dagger$ have the following properties:

$$a^\dagger |j\rangle = \sqrt{j+1} |j+1\rangle ,$$
$$a|0\rangle = 0 \quad \text{and} \quad a|j\rangle = \sqrt{j} |j-1\rangle \quad \text{for} \quad j > 0 . \tag{2.5}$$

Consequently, the matrix representation of q in the space of the unperturbed states $|j\rangle$ yields the tridiagonal matrix

$$Q_{jk} = \langle j|q|k\rangle$$
$$= \frac{1}{\sqrt{2}} \sqrt{k+1} \delta_{j,k+1} + \frac{1}{\sqrt{2}} \sqrt{k} \delta_{j,k-1} = \frac{1}{2} \sqrt{j+k+1} \delta_{|k-j|,1} . \tag{2.6}$$

The approximation that we will now use consists of defining this infinite-dimensional matrix for $j, k = 0, 1, \ldots, n-1$ only. The Hamiltonian $H = H_0 + \lambda q^4$, or more precisely its matrix representation, is to be approximated by an $n \times n$ matrix as well, where H_0 is represented by the diagonal matrix with the elements $j + 1/2$, and q^4 by the fourfold matrix product of Q_{jk} with itself. The error of the approximation can be estimated by comparing the eigenvalues of H for different values of n.

Algorithm

The above matrices can be defined in a particularly compact way in *Mathematica*. According to (2.6), $\langle j|q|k\rangle$ is

```
q[j_,k_]:= Sqrt[(j+k+1)]/2 /; Abs[j-k]==1
q[j_,k_]:= 0 /; Abs[j-k] != 1
```

The construct `lhs := rhs /; test` means that the definition is to be used only if the expression on the right hand side has the logical value `True`. The matrix q is defined as a list of lists

```
q[n_]:= Table[q[j,k], {j,0,n-1}, {k,0,n-1}]
```

and according to (2.2) H_0 is calculated as

```
h0[n_]:= DiagonalMatrix[Table[j+1/2, {j,0,n-1}]]
```

With this, H can be written as

```
h[n_]:= h0[n] + lambda q[n].q[n].q[n].q[n]
```

One can calculate the eigenvalues of H with `Eigenvalues[...]`, either algebraically (which only works for small values of n) or numerically with `Eigenvalues[N[h[n]]]`, where `lambda` has to be assigned a numerical value beforehand.

Results

The call `h[4] // MatrixForm` yields the Hamiltonian matrix

$$
\begin{pmatrix}
\frac{1}{2}+\frac{3}{4}\lambda & 0 & \frac{3}{\sqrt{2}}\lambda & 0 \\
0 & \frac{3}{2}+\frac{15}{4}\lambda & 0 & 3\sqrt{\frac{3}{2}}\lambda \\
\frac{3}{\sqrt{2}}\lambda & 0 & \frac{5}{2}+\frac{27}{4}\lambda & 0 \\
0 & 3\sqrt{\frac{3}{2}}\lambda & 0 & \frac{7}{2}+\frac{15}{4}\lambda
\end{pmatrix}.
\tag{2.7}
$$

Its eigenvalues are

$$
\frac{6+15\lambda\pm2\sqrt{2}\sqrt{2+12\lambda+27\lambda^2}}{4} \quad\text{and}\quad \frac{10+15\lambda\pm2\sqrt{2}\,\sqrt{2+27\lambda^2}}{4}.
\tag{2.8}
$$

These four eigenvalues can be seen in Fig. 2.1 as functions of λ. Without perturbation ($\lambda = 0$) we get the energies 1/2, 3/2, 5/2, and 7/2 of the harmonic oscillator. As the perturbation λ increases, the upper two eigenvalues separate markedly from the lower ones. This is an effect of the finite size of the matrices. The operator q^4 connects the state $|j\rangle$ with $|j\pm4\rangle$, and therefore the matrix elements $\langle j|q^4|j\pm4\rangle$ should be contained in the Hamiltonian matrix $h[n]$, if the energy ε_j is to be calculated more or less correctly. In any case, it is no surprise if the eigenvalues ε_{n-1}, ε_{n-2}, and ε_{n-3} exhibit large discrepancies from the exact result.

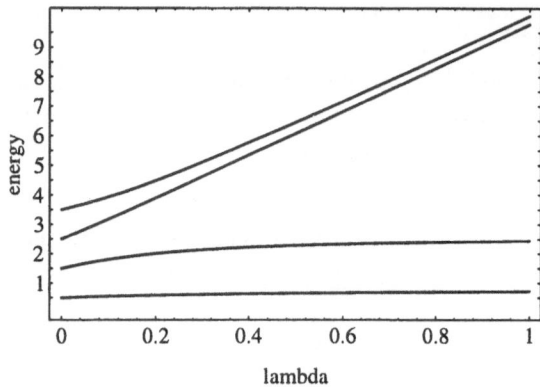

Fig. 2.1. Approximate values for the lowest four energy eigenvalues of the quantum oscillator with the anharmonicity λq^4, as a function of λ

How accurate is the approximation? This can be seen in Fig. 2.2 for the example of the ground state energy ε_0. For an anharmonicity parameter $\lambda = 0.1$, ε_0 is plotted as a function of $1/n$, for $n = 7, \ldots, 20$. Obviously this function is not monotonic. Although we do not know its asymptotic behavior, the value for $n \to \infty$ can be specified very precisely. By using the commands

```
mat = N[h[n] /. lambda -> 1/10, 20];
li = Sort[Eigenvalues[mat]]; li[[1]]
```

we obtain the following values for $n = 20$ and $n = 40$:

$$\varepsilon_0\,(20) = 0.559146327396\ldots,$$
$$\varepsilon_0\,(40) = 0.559146327183\ldots.$$

Therefore, for $n = 20$ we have an accuracy of about nine significant figures. Higher energies can only be determined with less accuracy. We obtain

$$\varepsilon_{10}\,(20) = 17.333\ldots,$$
$$\varepsilon_{10}\,(40) = 17.351\ldots,$$

i.e., an accuracy of only three significant figures.

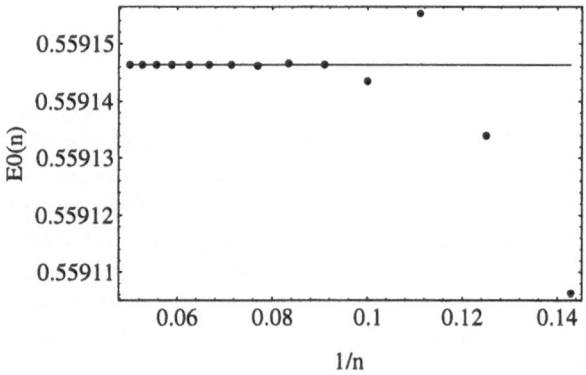

Fig. 2.2. Approximate values for the ground state energy of the anharmonic oscillator as a function of the inverse matrix dimension $1/n$ for $n = 7, 8, \ldots, 20$

In Fig. 2.1 we only considered four levels for our approximation. Therefore, the two upper energies are represented in an entirely incorrect manner. In Fig. 2.3, on the other hand, we have included twenty levels. In this case, the numerical solution of the eigenvalue equation yields a very precise result for the five lowest energies. The energies, as well as their separations, increase as the anharmonicity parameter λ increases.

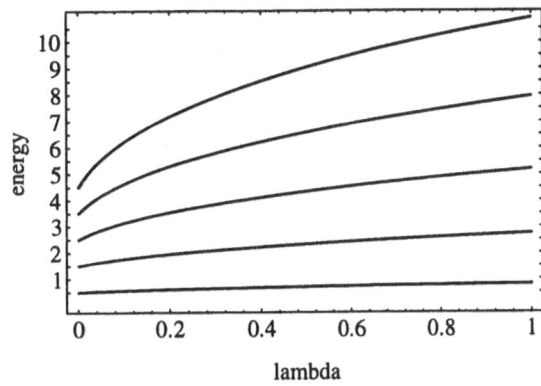

Fig. 2.3. The five lowest energy eigenvalues as a function of the anharmonicity parameter λ

Exercises

We consider a quantum particle in one dimension in a double-well potential. For the Hamiltonian, we use the dimensionless form

$$H = \frac{p^2}{2} - 2q^2 + \frac{q^4}{10} .$$

1. Draw the potential.
2. Calculate the four lowest energy levels.
3. Plot the wave functions of the four lowest levels together with the potential.
4. One can also calculate the matrix elements of q^2 and q^4 directly by expressing the operators in terms of a and $a^\dagger$, analogously to (2.4). Does this improve the results?

Literature

Schwabl F. (1995) Quantum Mechanics. Springer, Berlin, Heidelberg, New York

2.2 Electrical Circuits

Ohm's law is a linear relation between current and voltage. It is still true for monofrequency alternating currents and alternating voltages, if capacitors and inductive coils are represented by complex impedances. The equations even remain valid for the general case of passive electrical networks, if one chooses an appropriate representation for currents and voltages in complex space. For a given frequency, they can be solved relatively easily by computer, and with the Fourier transformation, the output voltage can then be calculated for every input signal.

Physics

We consider an alternating voltage $V(t)$ and the corresponding alternating current $I(t)$ and express both as complex-valued functions:

$$V(t) = V_0 e^{i\omega t} ,$$
$$I(t) = I_0 e^{i\omega t} , \tag{2.9}$$

where V_0 and I_0 are complex quantities whose phase difference indicates how much the current oscillation precedes or lags behind the voltage oscillation. Strictly speaking, only the real part of (2.9) has a physical meaning, but the phase relations are especially easy to formulate in complex notation. With this, Ohm's law becomes

$$V_0 = Z I_0 \tag{2.10}$$

with a complex impedance Z. For an ohmic resistance R, for a capacitance C, and for an inductance L, Z is given by

$$Z = R , \quad Z = \frac{1}{i\omega C} , \quad Z = i\omega L , \tag{2.11}$$

where R, C, and L are real quantities, measured for example in the units ohm, farad, and henry. In an electric network, the following conservation laws, also known as Kirchhoff's laws, are valid:

1. Owing to charge conservation, the sum of the incoming currents is equal to the sum of the outgoing currents at every node.
2. Along any path, the partial voltage drops across each element add up to the total voltage over the path.

Together with Ohm's law, these two conditions yield a system of equations which determines all unknown currents and voltages.

As a simple example, we consider an L–C oscillatory circuit which is connected in series with a resistance R (Fig. 2.4). Let V_i and V_o be the complex amplitudes of the input and output voltages with the angular frequency ω, and let I_R, I_C, and I_L be the amplitudes of the currents, which, after the

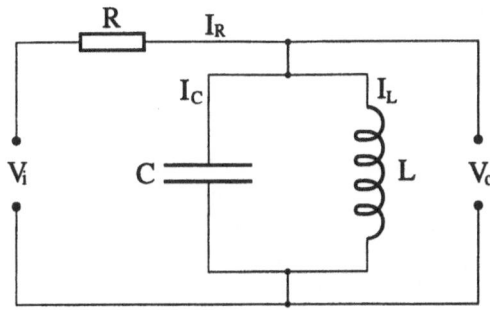

Fig. 2.4. Series connection of resistance and L–C oscillatory circuit

transient, have the same frequency as the input voltage. In this case, the following equations hold:

$$\text{Voltage addition: } V_R + V_o = V_i \, ,$$
$$\text{Current conservation: } I_R = I_C + I_L \, ,$$
$$\text{Ohm's law: } V_R = R\, I_R \, , \tag{2.12}$$
$$V_o = \frac{1}{i\omega C}\, I_C \, ,$$
$$V_o = i\omega L\, I_L \, .$$

For a given input voltage V_i, these five equations, which in this simple case are easily solved without a computer, determine the five unknowns V_R, V_o, I_R, I_C, and I_L. Independently of R, the magnitude of the output voltage V_o always reaches a maximum at $\omega = 1/\sqrt{LC}$; at this frequency the impedance of the oscillatory circuit is infinite.

In Fig. 2.5 we have expanded the network by adding a series circuit. If a capacitor C and an inductance L are connected in series, then the impedance at the frequency $\omega = 1/\sqrt{LC}$ is minimal. Consequently, we expect a maximal output voltage for this circuit at this frequency. This network is described by the following two equations:

$$\text{Voltage addition: } I_R \left(R + \frac{1}{i\omega C} + i\omega L \right) + V_o = V_i \, ,$$

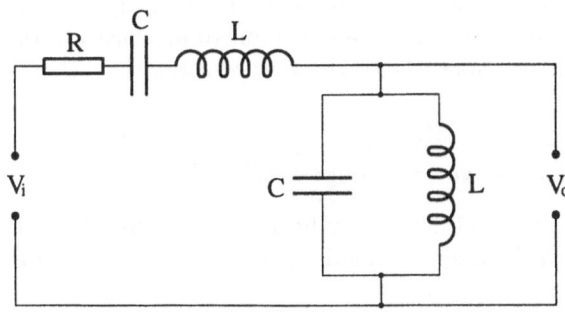

Fig. 2.5. Series connection of R–C–L combination and L–C oscillatory circuit

Current conservation: $I_R = \left(i\omega C + \dfrac{1}{i\omega L} \right) V_o$,   (2.13)

where Ohm's law has already been inserted. Surprisingly, the result for $V_o(\omega)$ is entirely different from what we (not being experts on electronics) expected. Two new resonances appear above and below $\omega = 1/\sqrt{LC}$; there, for small resistances R, V_o becomes much larger than V_i. This can be made plausible by the following consideration: For low frequencies, the behavior of the parallel circuit is dominated by the inductance. If we completely neglect the capacitor in the parallel circuit for the time being, then we are dealing with a series connection consisting of the elements R–C–L–L. Because the overall inductance is now $2L$, we get a resonance at $\omega_1 = 1/\sqrt{2LC}$. If, on the other hand, for high frequencies only the capacitance in the parallel circuit is considered, then the result is an R–C–L–C series connection with a total capacitance of $C/2$ and a corresponding resonance frequency $\omega_2 = \sqrt{2/(LC)}$. In the results section we will compare this approximation with the exact results.

Up to this point, we have only considered monofrequency sinusoidal voltages and currents. Now we want to apply an arbitrary periodic input signal $V_i(t)$ to the circuit. Because the network is linear, from a superposition of input signals one obtains the corresponding superposition of the output voltages. In particular, one can expand any periodic input voltage $V_i(t)$ with period T into a Fourier series:

$$V_i(t) = \sum_{n=-\infty}^{\infty} V_i^{(n)} \exp\left(2\pi i n \frac{t}{T} \right) .$$ (2.14)

For every term with the amplitude $V_i^{(n)}$ one can use (2.13) for the frequency $\omega_n = 2\pi n/T$ to obtain an output voltage $V_o(\omega_n) = V_o^{(n)}$, so that the total output signal is given by

$$V_o(t) = \sum_{n=-\infty}^{\infty} V_o^{(n)} \exp\left(2\pi i n \frac{t}{T} \right) .$$ (2.15)

Algorithm

In *Mathematica*, (2.12) and (2.13) can be entered directly. Although one can immediately solve both systems manually, we still want to demonstrate the principle using these simple examples. With the normalization $V_i = 1$, the system (2.12) becomes

```
eq1 = {vr + vo == 1, ir == ic + il, vr == ir r,
       vo == ic/(I omega c), vo == I omega L il}
```

One should note that the first equals sign, =, indicates an assignment, while == yields a logical expression. Consequently, a list of equations is assigned to the variable eq1. With

```
Solve[eq1, {vo, vr, ir, ic, il}]
```

the system of equations is solved for the specified variables. Because systems of equations generally have several solutions, Solve returns a list with lists of rules. Since there is only one solution in this case, though, we extract the first – and in this case the only one – of them via Solve[...][[1]].

As an example of a non-sinusoidal input voltage $V_i(t)$ we choose a saw-tooth voltage with the period T, which we probe at N equidistant points in time, in order to be able to use the discrete Fourier transformation. Thus we define discrete voltage values by

$$a_r = V_i(t_r) \equiv V_i\left(\frac{(r-1)T}{N}\right) \ , \quad r = 1, \ldots, N \ ,$$

and by using the inverse Fourier transformation we obtain the coefficients b_s with the property

$$a_r = \frac{1}{\sqrt{N}} \sum_{s=1}^{N} b_s \exp\left[2\pi i \frac{(s-1)(r-1)}{N}\right]$$

or

$$V_i(t_r) = \frac{1}{\sqrt{N}} \sum_{s=1}^{N} b_s \exp\left[2\pi i \frac{(s-1)t_r}{T}\right] \ .$$

We cannot use the Fourier transformation itself here, but have to use the inverse transformation instead, so that the sign in the argument of the exponential function agrees with (2.9). Although the authors of *Mathematica* declared that they wanted to follow the physicists' convention as far as this choice of sign in the Fourier transformation is concerned, they have realized exactly the opposite.

The amplitude $b_s/\sqrt{N}$ thus belongs to the frequency $\omega_s = 2\pi(s-1)/T$. At the output, every amplitude b_s is multiplied by the output voltage $V_o(\omega_s)$, which we obtained from the above equations, by using Solve[...]. This is only valid for $s = 1, \ldots, N/2$, however. Higher frequencies result in an inaccurate approximation for $V_i(t)$, as shown in detail in Sect. 1.3. One must shift the higher frequencies to low negative ones, using $b_s = b_{s-N}$, before transforming the $V_o(\omega_s)$. The transformed Fourier coefficients are

$$b_s^t = b_s \, V_o(\omega_s) \ , \quad s = 1, \ldots, \frac{N}{2} \ ,$$

$$b_s^t = b_s \, V_o(\omega_{s-N}) \ , \quad s = \frac{N}{2} + 1, \ldots, N \ .$$

The inverse transformation, which in this case is the Fourier transformation itself, then yields the output signal.

Results

The solution of the system of equations (2.12), which describes the series connection (shown in Fig. 2.4) of ohmic resistance and L–C oscillatory circuit, is contained in the variable **vos** in the *Mathematica* program. A readable form of the solution is

$$V_o(\omega) = \frac{-i\omega L}{-i\omega L - R + RCL\omega^2} \; .$$

Obviously, the network has a resonance at $\omega_r = 1/\sqrt{LC}$, as can also be seen in Fig. 2.6 for different values of R.

We have chosen $L = 1$ mH and $C = 1$ µF, which yields a resonance frequency ω_r of $31\,622.8$ s^{-1}. For $R = 0$, one obtains $V_o(\omega) = 1$, i.e., no resonance, whereas for $R \to \infty$, a sharp voltage maximum appears at the position of the resonance frequency. At ω_r the phase of the output voltage changes from $+\pi/2$ to $-\pi/2$. If we now apply a sawtooth voltage $V_i(t)$ to this filter, then the result will, of course, depend on the sharpness of the resonance at ω_r, and on the ratio of the fundamental frequency of the sawtooth voltage to the resonance frequency. Therefore, we parametrize the period T of $V_i(t)$ in the form $2\pi/T = f\omega_r$; f thus specifies the ratio of the input fundamental frequency to the resonance frequency. For $f = 1$ and a correspondingly narrow resonance curve the filter should convert the sawtooth-shaped oscillation to just a sinusoidal oscillation with the frequency $\omega = \omega_r$.

Basically, for $f < 1/2$ and $R = 200\,\Omega$, for example, only the multiple of $2\pi/T$ closest to ω_r will be filtered out. By contrast, for $f > 1$ and a broad resonance, all harmonics of $2\pi/T$ are included and a distorted copy of the input signal appears. This expectation is confirmed in Figs. 2.7 and 2.8.

For the second example, the series connection of R–C–L combination and L–C oscillatory circuit (Fig. 2.5), *Mathematica*, again with $V_i = 1$, gives the solution

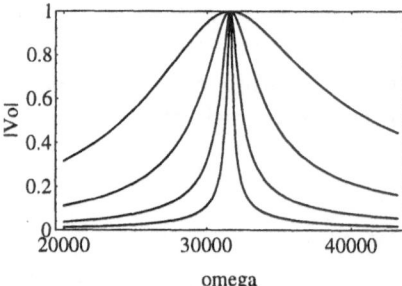 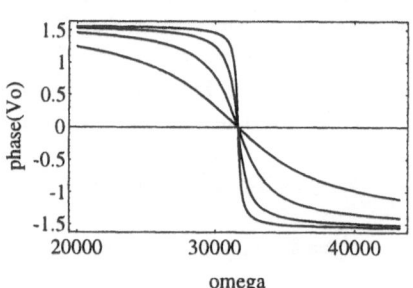

Fig. 2.6. Frequency dependence of the magnitude (*left*) and phase (*right*) of the output voltage $V_o(\omega)$ for the network shown in Fig. 2.4. The curves correspond to the resistances $R = 100\,\Omega$, $300\,\Omega$, $900\,\Omega$, and $2700\,\Omega$. The larger the resistance, the sharper the resonance

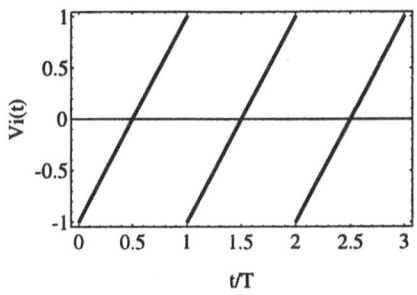
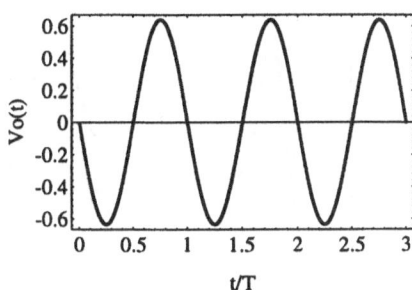

Fig. 2.7. The original sawtooth voltage $V_i(t)$ (*left*) and the voltage $V_o(t)$ at the output of the filter for $f = 1$ and $R = 2700\,\Omega$ (*right*)

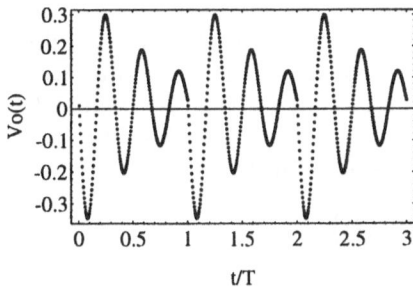
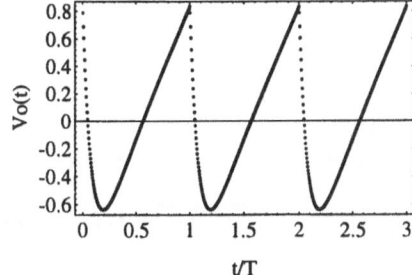

Fig. 2.8. The output voltage for $f = 1/3$ and $R = 200\,\Omega$ (*left*), and the distorted copy of the input signal that results for $f = 3$ and $R = 5\,\Omega$ (*right*)

$$V_o = \frac{CL\omega^2}{CL\omega^2 + (CL\omega^2 - 1)(1 - CL\omega^2 + iCR\omega)} .$$

For $R = 0$, i.e., if the ohmic resistance is equal to zero and the circuit is therefore loss-free, the denominator vanishes at

$$\omega = \sqrt{\frac{3 \pm \sqrt{5}}{2LC}} .$$

Then, with the values $L = 1$ mH and $C = 1$ μF, V_o diverges at the frequencies $\omega = 19\,544$ s^{-1} and $\omega = 51\,166.7$ s^{-1}. We can see that our previous estimate for the two resonances, $\omega_1 = 1/\sqrt{2LC} = 22\,360$ s^{-1} and $\omega_2 = \sqrt{2/(LC)} = 44\,720$ s^{-1}, was not entirely incorrect.

Figure 2.9 shows $V_o(\omega)$ for $R = 10$, 30, and 90 Ω. For any value of the resistance R, the output voltage at the frequency $\omega = 1/\sqrt{LC}$ is equal to the input signal, $V_o = V_i$. The two resonances are visible only if the resistances are small.

In closing, we want to calculate the power that is converted to heat by the resistor R at the frequency ω. Here, too, the complex representation of current and voltage proves to be advantageous. The calculation – done at first for a general element with a complex impedance Z, through which a

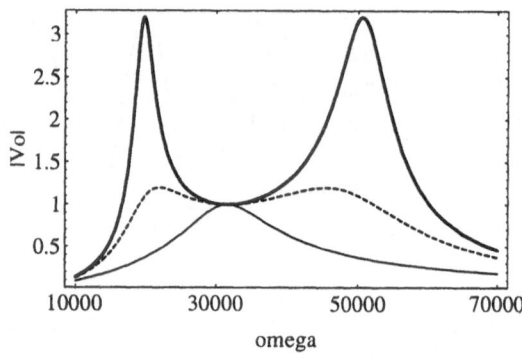

Fig. 2.9. Magnitude of the output voltage for the network of series and parallel circuit. The corresponding values of the resistances are (from top to bottom) $R = 10\,\Omega$, $30\,\Omega$, and $90\,\Omega$

current $I(t) = I_Z \exp(i\omega t)$ flows and across which there is a voltage drop $V(t) = V_Z \exp(i\omega t)$ – goes as follows: the power P is the time average of the product of the real parts of current and voltage, $P = \overline{\mathrm{Re}\, I(t)\mathrm{Re}\, V(t)}$. Owing to the purely harmonic time dependence, however, we have $\mathrm{Re}\, I(t) = \mathrm{Im}\, I(t + \pi/(2\omega))$, and correspondingly for $V(t)$. Thus we obtain

$$P = \frac{1}{2}\left(\overline{\mathrm{Re}\, I(t)\,\mathrm{Re}\, V(t)} + \overline{\mathrm{Im}\, I(t)\,\mathrm{Im}\, V(t)}\right)$$

$$= \frac{1}{4}\left(\overline{I(t)\, V(t)^*} + \overline{I(t)^*\, V(t)}\right)$$

$$= \frac{1}{4}\left(I_Z V_Z^* + I_Z^* V_Z\right)$$

$$= \frac{1}{2}|I_Z|^2\, \mathrm{Re}\, Z\,, \tag{2.16}$$

where we have set $V_Z = Z I_Z$ at the end. For the ohmic resistance R, the power is accordingly $P(\omega) = |I_R(\omega)|^2 R/2$. We compare this to the power P_0, which results if all coils and capacitors in our network are short-circuited, i.e., if the input voltage V_i is connected directly to the resistor R, so that one gets $I_R = V_i/R$, and therefore $P_0 = |V_i|^2/(2R)$. The power ratio $P(\omega)/P_0$ is shown in Fig. 2.10 for $R = 10\,\Omega$. At both resonance frequencies the coils and capacitors appear to be completely conductive, while at $\omega = 1/\sqrt{LC}$ the parallel circuit offers an infinite resistance.

Exercises

In the second R–C–L network (Fig. 2.5), add a resistor R to the parallel L–C circuit, parallel to the capacitor and the coil.

1. Calculate and draw $V_0(\omega)$.
2. What form does a periodic rectangular voltage have after passing through this filter?
3. What is the power dissipation in the two resistances as a function of ω?

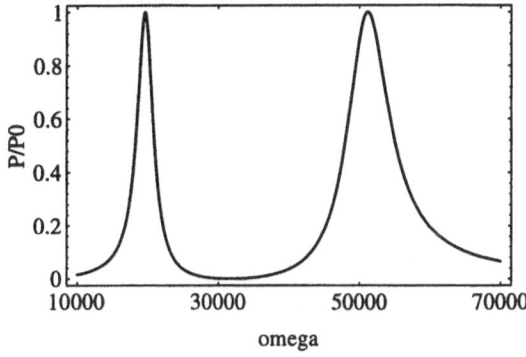

Fig. 2.10. The power $P(\omega)$ dissipated in the ohmic resistance $R = 10\,\Omega$, relative to the power P_0 which is dissipated if the input voltage is connected directly to R

Literature

Crandall R.E. (1991) Mathematica for the Sciences. Addison-Wesley, Redwood City, CA

2.3 Chain Vibrations

It is a well known fact that the motion of a particle in a quadratic potential is described by an especially simple linear differential equation. If several particles interact with each other through such linear forces, their motion can be calculated by linear equations as well. However, one then has several such equations of motion which are coupled to each other. This system of linear equations can be solved by diagonalizing a matrix. A good example of this is the linear chain with different masses. It is a simple model for the calculation of lattice vibrations in a crystal. The eigenvalues of a matrix specify the energy bands of the phonons, while the eigenvectors provide information about the vibration modes of the crystal elements. Every possible motion of the model solid can be represented by superposition of such eigenmodes.

Physics

We consider a chain consisting of pointlike masses m_1 and m_2, which we designate as *light* and *heavy* atoms respectively, for sake of simplicity. The particles are to be arranged in such a way that one heavy atom follows three light ones. The unit cell of length a thus contains four atoms. Only nearest neighbors shall interact with one another. We limit our considerations to small displacements, i.e., the forces are to be linear functions of the shifts of the masses, as indicated in the spring model shown in Fig 2.11.

To describe the longitudinal oscillations, we number the unit cells sequentially and consider the cell with the number n. Within this cell, let r_n, s_n, and t_n be the displacements of the light atoms from their rest positions, and

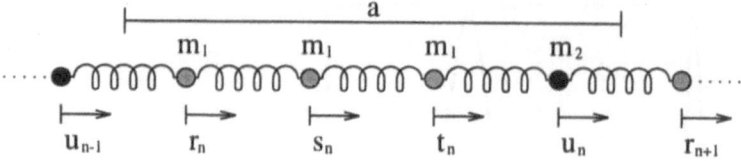

Fig. 2.11. Linear chain consisting of two types of atoms which are connected to each other by elastic forces

let u_n be the displacement of the heavy atom from its rest position. Then the equations of motion are

$$
\begin{aligned}
m_1 \ddot{r}_n &= f\left(s_n - r_n\right) - f\left(r_n - u_{n-1}\right) \\
&= f\left(s_n + u_{n-1} - 2r_n\right) , \\
m_1 \ddot{s}_n &= f\left(t_n + r_n - 2s_n\right) , \\
m_1 \ddot{t}_n &= f\left(u_n + s_n - 2t_n\right) , \\
m_2 \ddot{u}_n &= f\left(r_{n+1} + t_n - 2u_n\right) .
\end{aligned}
\tag{2.17}
$$

Here, f is the spring constant. In an infinitely long chain, these equations hold for every unit cell $n \in \mathbb{Z}$. For a finite chain consisting of N unit cells we assume periodic boundary conditions, that is, we think of it as a ringlike arrangement of N identical cells. Since the energy of the chain does not change if it is shifted by the distance a, and consequently the equations of motion are invariant under the translations $\{r_n, s_n, t_n, u_n\} \to \{r_{n+k}, s_{n+k}, t_{n+k}, u_{n+k}\}$, $k = 1, 2, \ldots, N$, (2.17) can be solved by applying a Fourier transformation. Therefore we use the ansatz

$$
\boldsymbol{x}_n(t) = \begin{pmatrix} r_n(t) \\ s_n(t) \\ t_n(t) \\ u_n(t) \end{pmatrix} = \boldsymbol{S}(q) \exp\left(iqan \pm i\omega t\right) ,
\tag{2.18}
$$

where, owing to the periodic boundary conditions, q can only take the values $q_\nu = 2\pi\nu/(Na)$, $\nu = -N/2 + 1, \ldots, N/2$. Then

$$
\ddot{\boldsymbol{x}}_n = -\omega^2 \boldsymbol{x}_n, \quad \boldsymbol{x}_{n\pm1} = \exp\left(\pm iqa\right)\boldsymbol{x}_n .
\tag{2.19}
$$

This inserted into (2.17) yields

$$
\begin{pmatrix} 2f & -f & 0 & -fe^{-iqa} \\ -f & 2f & -f & 0 \\ 0 & -f & 2f & -f \\ -fe^{iqa} & 0 & -f & 2f \end{pmatrix} \boldsymbol{S}(q) = \omega^2 \begin{pmatrix} m_1 & 0 & 0 & 0 \\ 0 & m_1 & 0 & 0 \\ 0 & 0 & m_1 & 0 \\ 0 & 0 & 0 & m_2 \end{pmatrix} \boldsymbol{S}(q) ,
\tag{2.20}
$$

which is a generalized eigenvalue equation of the type

$$
\mathsf{F}\boldsymbol{S} = \lambda \mathsf{M} \boldsymbol{S} .
\tag{2.21}
$$

Here, M is the mass matrix, and we are looking for the eigenvalues $\lambda(q) = \omega^2(q)$ and the corresponding normal modes $\boldsymbol{S}(q)$ for a given value of q. In our case, the mass matrix can easily be inverted and (2.21) is obviously equivalent to

$$\mathsf{M}^{-1}\mathsf{F}S = \lambda S \ . \tag{2.22}$$

$\mathsf{M}^{-1}\mathsf{F}$ is not a Hermitian matrix, however. The following transformation of (2.21) shows that the eigenvalues are real and nonnegative in spite of this. We multiply this equation from the left by $\mathsf{M}^{-1/2}$, the inverse of the matrix $\mathsf{M}^{1/2}$, and obtain

$$\mathsf{M}^{-1/2}\mathsf{F}\mathsf{M}^{-1/2}\mathsf{M}^{1/2}S = \lambda \mathsf{M}^{1/2}S \ . \tag{2.23}$$

This is an ordinary eigenvalue equation for the positive semidefinite Hermitian matrix $\mathsf{M}^{-1/2}\mathsf{F}\mathsf{M}^{-1/2}$ with the eigenvector $\mathsf{M}^{1/2}S$.

The eigenmodes $x_n(t) = S_\ell(q_\nu) \exp\left(iq_\nu an \pm i\omega_{\nu\ell}t\right)$ thus obtained are particular complex-valued solutions of the equations of motion (2.17). Analogously to the electrical filters (Sect. 2.2), the general solution can be obtained by a superposition of the eigenmodes. From the complex conjugate of (2.20) and the substitution $q \to -q$ we see that, along with $S(q)$, $S(-q)^*$ is an eigenvector as well, with the same eigenvalue. Therefore, the general real solution of (2.17) has the form

$$x_n(t) = \sum_{\nu,\ell} S_\ell(q_\nu) \exp\left(iq_\nu an\right)$$
$$\times \left[c_{\nu\ell} \exp\left(i\omega_{\nu\ell}t\right) + c^*_{-\nu\ell} \exp\left(-i\omega_{\nu\ell}t\right)\right] \ . \tag{2.24}$$

The coefficients $c_{\nu\ell}$, which are not yet determined, are fixed by the initial conditions.

Algorithm and Results

The 4×4 matrix (2.22) does not represent any particular challenge to an analytical solution of the eigenvalue equation. If we consider several different types of atoms, or the corresponding two-dimensional problem, however, the matrices become so large that only a numerical determination of the vibration modes is possible. We want to demonstrate the latter using our simple example.

First, we write the matrices F and M in a form suited for *Mathematica*, choosing $a = 1$:

```
mat1 = { {        2f, -f,   0, -f*Exp[-I q]},
         {        -f, 2f,  -f,           0},
         {         0, -f,  2f,          -f},
         {-f*Exp[I q], 0,  -f,          2f} }

massmat = DiagonalMatrix[{m1, m1, m1, m2}]
```

From this we form the product mat2 = Inverse[massmat].mat1. In this case, *Mathematica* is still capable of specifying the eigenvalues in a general form in response to the command Eigenvalues[mat2]. The calculation is complex, however, and the result consists of nested square and cube roots; therefore we prefer a numerical calculation.

```
eigenlist = Table[{x, Chop[Eigenvalues[
            mat2 /.{f->1.,m1->0.4,m2->1.0,q->x}]]},
            {x,-Pi,Pi,Pi/50}]
```

In a short time this command yields a list of q values and the four corresponding squares of the frequencies.

```
Flatten[ Table[
         Map[{#[[1]],Sqrt[#[[2,k]]]}&, eigenlist],
         {k,4}], 1]
```

turns this result into a list of $(q-\omega)$ pairs, which were plotted with ListPlot in Fig. 2.12. As we can see, the allowed frequencies of the lattice vibrations form four bands. The lowest branch represents the so-called acoustic phonons, for which adjacent atoms almost always oscillate in the same direction. For $q = 0$, the entire chain is just shifted as a whole, which does not require any energy ($\omega = 0$). The three upper branches, the optical phonons, have many atoms oscillating against each other which requires energy even at $q = 0$.

The individual displacements can be determined from the eigenvectors of the matrix $M^{-1}F$. Figure 2.13 shows the four eigenmodes for $q = 0$. Surpris-

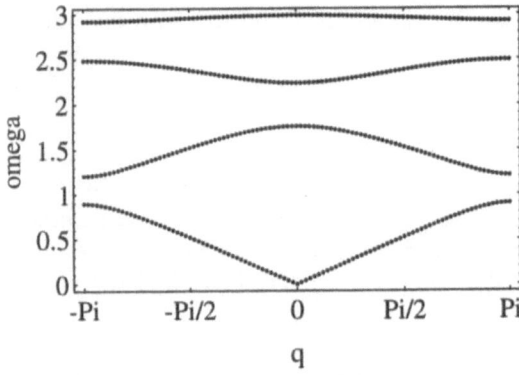

Fig. 2.12. The frequencies of the four eigenmodes of the linear chain from Fig. 2.11 as a function of the wave number q

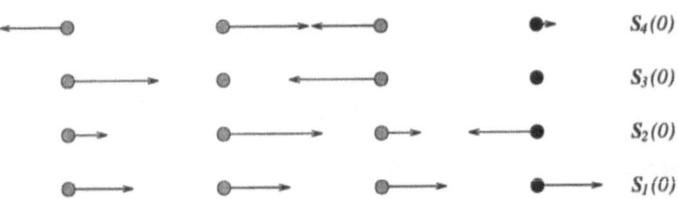

Fig. 2.13. Eigenmodes of the chain for $q = 0$

ingly the heavy atom and one of the light ones are at rest in the third branch, while the two other masses oscillate against each other.

Exercise

Investigate the two-dimensional vibrations of your single-family home. Five equal masses m are coupled by springs, and the potential between neighboring masses is

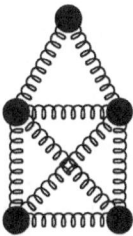

$$V\left(r_i, r_j\right) = \frac{D}{2}\left(|r_i - r_j| - a_{ij}\right)^2 ,$$

where the a_{ij} are the distances between the masses at rest, i.e., $a_{ij} = l$ for the sides and the roof, and $a_{ij} = \sqrt{2}l$ for the two diagonals.

Calculate the frequencies of the vibrations for small displacements and plot the eigenmodes. It is especially impressive if you produce an animation of the eigenmodes.

Literature

Goldstein H. (1980) Classical Mechanics. Addison-Wesley, Reading, MA
Kittel C. (1996) Introduction to Solid State Physics. Wiley, New York

2.4 The Hofstadter Butterfly

Linear equations can have intriguing solutions. This becomes especially apparent in the case of a crystal electron in a homogeneous magnetic field. The electron's energy spectrum as a function of the strength of the magnetic field has a complicated structure reminiscent of a butterfly. This problem was investigated in 1976 by Douglas Hofstadter. Amazingly, the differences between rational and irrational numbers become evident in this spectrum. We want to derive the quantum-mechanical equation for the energies of the electron and solve it numerically. A small computer program produces the intriguing butterfly.

Physics

We model an electron on a square lattice by localized wave functions and matrix elements for jumps between nearest neighbors. In such a *tight-binding* approximation the single-particle energies have the form

$$\varepsilon\left(k\right) = \varepsilon_0\left(\cos k_x a + \cos k_y a\right) , \qquad (2.25)$$

where a is the lattice constant, $4\varepsilon_0$ is the width of the energy band, and $k = (k_x, k_y)$ is the wave vector of an electron. The magnetic field is taken into account via the *Peierls trick*: $\hbar k$ is replaced with $p + eA/c$, where A is the vector potential and p is the momentum operator. For a magnetic field of magnitude B perpendicular to the x–y plane one can choose $A = (0, Bx, 0)$. Thus one obtains the Schrödinger equation

$$E \varphi\left(x, y\right) = \left[\varepsilon_0 \cos\left(\frac{a p_x}{\hbar}\right) + \varepsilon_0 \cos\left(\frac{a p_y}{\hbar} + \frac{aeBx}{\hbar c}\right)\right] \varphi\left(x, y\right) . \qquad (2.26)$$

Owing to the identity $\cos \alpha = [\exp(i\alpha) + \exp(-i\alpha)]/2$, and because the term $\exp(i a p_x / \hbar)$ is the translation operator that effects a shift by one lattice constant a, one finds:

$$
\begin{aligned}
E\varphi\left(x, y\right) = \frac{1}{2}\varepsilon_0 \Big[& \varphi\left(x + a, y\right) + \varphi\left(x - a, y\right) \\
& + \exp\left(\frac{iaeBx}{\hbar c}\right) \varphi\left(x, y + a\right) \\
& + \exp\left(-\frac{iaeBx}{\hbar c}\right) \varphi\left(x, y - a\right) \Big] .
\end{aligned}
\qquad (2.27)
$$

This turns the continuous wave equation into a discrete equation which couples $\varphi(x, y)$ with the amplitudes at the four neighboring sites. For the y-dependence of the wave function we assume a plane wave,

$$\varphi\left(x, y\right) = \exp\left(\frac{i\nu y}{a}\right) \psi\left(\frac{x}{a}\right) , \qquad (2.28)$$

and with the dimensionless variables $m = x/a$ and $\sigma = a^2 eB/hc$ we finally obtain the *Harper equation* for $\psi_m \equiv \psi(m)$,

$$\psi_{m-1} + 2 \cos\left(2\pi m\sigma + \nu\right) \psi_m + \psi_{m+1} = E\psi_m . \qquad (2.29)$$

Here, the energy E is measured in units of $\varepsilon_0/2$. This discrete Schrödinger equation contains only the parameters ν and σ. The parameter ν determines the y-component of the momentum, while σ is the ratio of the magnetic flux $a^2 B$ through the unit cell to a flux quantum hc/e.

With this equation we have mapped a two-dimensional lattice electron onto a particle that jumps along a chain in a periodic cosine potential. The quantity ν is the phase of the potential which does not play an important role, while σ specifies the ratio of the potential's period to the lattice constant ($= 1$ in units of m). Here one can already see the difference between rational and irrational values of σ: if σ is a rational number, i.e., $\sigma = p/q$ with p and q integer and relatively prime, then the potential is commensurate with the lattice with a period q. In this case, according to the Bloch theorem, the wave function has the property

$$\psi_{m+q} = \exp\left(ikq\right)\psi_m \ . \tag{2.30}$$

The eigenstates and the corresponding energies are classified by k and ν. If one writes (2.29) in matrix form in the following manner,

$$\begin{pmatrix} \psi_{m+1} \\ \psi_m \end{pmatrix} = M_1\left(m, E\right)\begin{pmatrix} \psi_m \\ \psi_{m-1} \end{pmatrix} \tag{2.31}$$

with

$$M_1\left(m, E\right) = \begin{pmatrix} E - 2\cos\left(2\pi m\sigma + \nu\right) & -1 \\ 1 & 0 \end{pmatrix} , \tag{2.32}$$

then the q-fold iterated equation combined with (2.30), namely

$$\prod_{r=1}^{q} M_1\left(q - r, E\right)\begin{pmatrix} \psi_0 \\ \psi_{-1} \end{pmatrix} = \exp\left(ikq\right)\begin{pmatrix} \psi_0 \\ \psi_{-1} \end{pmatrix} , \tag{2.33}$$

immediately shows that $\exp(ikq)$ is an eigenvalue of the matrix $M_q(E) = \prod_{r=1}^{q} M_1(q - r, E)$. With $\det(M_q) = 1$, the second eigenvalue of M_q has to be $\exp(-ikq)$. From this, we obtain the trace of M_q,

$$\text{Trace } M_q\left(E\right) = 2\cos kq \ . \tag{2.34}$$

The trace of the matrix $M_q(E)$ is a qth-order polynomial in E. Correspondingly, (2.34) has at most q real solutions $E(k, \nu)$. Therefore, we conclude that for rational numbers $\sigma = p/q$ there are at most q energy bands which one can calculate, in principle, from (2.34). In the vicinity of any rational number there are numbers with arbitrarily large q as well as irrational numbers ($q \to \infty$), however. Consequently, as a function of σ, there are spectra with large numbers of energy bands next to each spectrum with few bands. Here, the difference between rational and irrational numbers becomes evident in physics.

Algorithm

The numerical evaluation of the trace condition (2.34), which amounts to the determination of the zeros of a qth-order polynomial, turns out to be quite difficult. Therefore we choose a slightly different method which avoids this problem. For specific values of k and ν, namely for $k = 0$ and $\nu = 0$, we want to solve the Harper equation (2.29) numerically and to represent the result graphically as a function of the rational values $\sigma = p/q$. For this purpose, we let q loop over all even values between 4 and q_{max} and let p vary between 1 and $q - 1$, but select only those p for which p and q are relatively prime.

The reason for selecting only even q values is that the odd and the even cases have to be discussed separately; for the sake of simplicity we choose only one. Since $q = 2$ is a special case for which the general equations do not hold, the loop starts with $q = 4$. We take advantage of the fact that the potential term $V_m = 2\cos(2\pi mp/q)$ is an even function of m, so that the eigenvectors $\{\psi_m\}_{m=0}^{q-1}$ of the Harper equation can be assumed to be either odd or even functions of m. Indeed, the periodicity $\psi_{m+q} = \psi_m$ reduces the eigenvalue problem (2.29) to q independent components. Owing to the odd–even symmetry mentioned above, the problem can be further reduced to two eigenvalue problems of about half the size.

The assumption that $\{\psi_m\}$ is, for example, an odd function of m, that is $\psi_{-m} = -\psi_m$, first leads us to $\psi_0 = 0$ and then, taken together with the periodicity, to

$$\psi_{q/2+r} = -\psi_{-q/2-r} = -\psi_{q/2-r} \quad \text{for } r = 0, 1, \ldots, \frac{q}{2} - 1 . \qquad (2.35)$$

In particular, we also have $\psi_{q/2} = 0$, and for the remaining $q/2 - 1$ independent components $(\psi_1, \psi_2, \ldots, \psi_{q/2-1})^{\mathrm{T}} \equiv \boldsymbol{\psi}_o$, (2.29) can be written in matrix form,

$$\mathbf{A}_o \boldsymbol{\psi}_o = E \boldsymbol{\psi}_o , \quad \text{where } \mathbf{A}_o = \begin{pmatrix} V_1 & 1 & 0 & \cdots & 0 & 0 \\ 1 & V_2 & 1 & \ddots & & 0 \\ 0 & 1 & \ddots & \ddots & \ddots & \vdots \\ \vdots & \ddots & \ddots & \ddots & 1 & 0 \\ 0 & & \ddots & 1 & V_{q/2-2} & 1 \\ 0 & 0 & \cdots & 0 & 1 & V_{q/2-1} \end{pmatrix} . (2.36)$$

Thus, determining the allowed energies for the Harper equation in this case means calculating the eigenvalues of the tridiagonal matrix $\mathbf{A}_o$. Before we discuss a special numerical method, which will turn out to be particularly appropriate here, we want to demonstrate briefly that the treatment of those wave functions that are even in m leads to a very similar problem.

Let us assume that $\psi_{-m} = \psi_m$ and $\psi_{m+q} = \psi_m$. First of all, this implies that $\psi_{-1} = \psi_1$, and, analogous to (2.35), we now obtain

$$\psi_{q/2+r} = \psi_{-q/2-r} = \psi_{q/2-r} \quad \text{for } r = 1, 2, \ldots, \frac{q}{2} - 1 . \qquad (2.37)$$

In particular, $\psi_{q/2+1} = \psi_{q/2-1}$. As a consequence, the matrix representation of the Harper equation (2.29) for the $q/2 + 1$ independent components $(\psi_0, \psi_1, \ldots, \psi_{q/2})^{\mathrm{T}} \equiv \boldsymbol{\psi}_e$ of the wave function has the form

$$A_e \psi_e = E \psi_e \, , \quad \text{where} \quad A_e = \begin{pmatrix} V_0 & 2 & 0 & \cdots & 0 & 0 \\ 1 & V_1 & 1 & \ddots & & 0 \\ 0 & 1 & \ddots & \ddots & \ddots & \vdots \\ \vdots & \ddots & \ddots & \ddots & 1 & 0 \\ 0 & & \ddots & 1 & V_{q/2-1} & 1 \\ 0 & 0 & \cdots & 0 & 2 & V_{q/2} \end{pmatrix} . \quad (2.38)$$

As in the case of the odd eigenfunctions ψ_m, we are left with the task of determining the eigenvalues of a tridiagonal matrix, in this case those of A_e.

The algorithm for calculating the characteristic polynomial of a tridiagonal matrix can be formulated in the following recursive manner. We consider, for example, the matrix $A_o - E1$, in which we take A_o from (2.36). More specifically, we consider the upper left $n \times n$ submatrix of that matrix. If we designate its determinant by $p_n(E)$, then an expansion in terms of the last row yields

$$p_n(E) = (V_n - E) p_{n-1}(E) - p_{n-2}(E) \, . \quad (2.39)$$

This two-part recursion formula is completed by the initial values

$$p_1(E) = V_1 - E \quad \text{and} \quad p_0(E) = 1 \, . \quad (2.40)$$

In this way, one can recursively calculate $\det(A_o - E1) = p_{q/2-1}(E)$.

The special importance of the polynomials $p_n(E)$, however, comes from the fact that they form a so-called *Sturm sequence*: Sturm sequences contain information about the zeros of the highest-order polynomial. If one considers the number $c(E)$ of sign changes which occur at the position E in the sequence

$$p_0(E), \ p_1(E), \ p_2(E), \ \ldots, \ p_{q/2-1}(E) \, ,$$

then one obtains the result:

The number of zeros of $p_{q/2-1}(E)$ in the interval $E_1 \leq E < E_2$ is equal to $c(E_2) - c(E_1)$.

Because there is always a finite resolution in a graphical representation, e.g., on the screen the pixel size, we will generate a suitable subdivision $E_0 < E_1 < E_2 \cdots < E_N$ of the accessible energy range. The numbers of sign changes $c(E_i)$ and $c(E_{i+1})$ then tell us whether there are any eigenvalues between E_i and E_{i+1}. The region in which any energy eigenvalues can be found at all can be narrowed down using Gershgorin's theorem: For every eigenvalue E of an $n \times n$ matrix $A = (a_{ij})$ there is an i for which

$$|E - a_{ii}| \leq \sum_{j=1, j \neq i}^{n} |a_{ij}| \, . \quad (2.41)$$

Because in our case we have $\sum_{j=1, j \neq i}^{n} |a_{ij}| \leq 2$ and $|a_{ii}| \equiv |V_i| \leq 2$, we conclude that the eigenvalues fall between -4 and 4.

For the matrix A_e, which differs slightly from A_o, the algorithm corresponding to (2.39) and (2.40) is

$$p_0(E) = 1 ,$$
$$p_1(E) = V_0 - E ,$$
$$p_2(E) = (V_1 - E)\, p_1(E) - 2p_0(E) , \qquad (2.42)$$
$$p_n(E) = (V_{n-1} - E)\, p_{n-1}(E) - p_{n-2}(E) \quad \text{for } n = 3, 4, \ldots, \frac{q}{2} ,$$
$$p_{q/2+1}(E) = (V_{q/2} - E)\, p_{q/2}(E) - 2p_{q/2-1}(E) .$$

We want to point out that the equations for the Sturm sequences, (2.39) and (2.42), are almost identical to the Harper equation itself, for (2.29) leads to the following relations for the odd wave functions $\psi_{-m} = -\psi_m$ with the initial values $\psi_0 = 0$, $\psi_1 = 1$:

$$\psi_2 = (E - V_1) , \quad \psi_3 = (E - V_2)\,\psi_2 - 1 , \quad \psi_4 = (E - V_3)\,\psi_3 - \psi_2 , \ldots$$

The choice $\psi_1 = 1$ is arbitrary; it fixes the normalization of the wave function but has no effect on the eigenvalue. $\psi_0 = 0$ follows from the symmetry. Comparison with (2.39) and (2.40) yields

$$\psi_1 = p_0(E) , \quad \psi_2 = -p_1(E) , \quad \psi_3 = p_2(E) , \quad \psi_4 = -p_3(E) , \quad \ldots .$$

Now if the number of sign changes, or nodes, in the sequence $\{p_n(E)\}$ changes, then the number of nodes in the sequence $\{\psi_n\}$ changes as well, by the same amount. Thus the number of eigenvalues in the interval $[E_1, E_2]$ is determined by the difference in the number of nodes between $\{\psi_n(E_1)\}$ and $\{\psi_n(E_2)\}$. This is reminiscent of the node theorem of one-dimensional quantum mechanics, which we will use again for the anharmonic oscillator in Sect. 4.3: The number of nodes of a stationary wave function $\psi(x)$ determines the number of the energy level above the ground state.

The algorithm above can be written in just a few lines of code. In Fig. 2.14 we have chosen to divide the complete energy range $|E| \leq 4$ into 500 bins. For a resolution of only 500 points on the horizontal axis the total number of computation steps can be estimated to be about $500 \times 500 \times 500 \times 10 \simeq 10^9$, where we have assumed ten operations for the innermost loop. Because *Mathematica* would be too slow for this, we have written the program in C. After the usual initialization, it essentially consists of four loops: the q loop and the p loop, where $\sigma(p, q)$ is varied, the ie loop, in which the energy changes between -4 and 4, and the inner loop over m, which is used to determine the number of sign changes in the sequence of polynomials $p_0, p_1, \ldots, p_{q/2\pm1}$. This part of the program has the following form:

```
for(q = 4; q < qmax; q+=2)
    {
    for(p = 1; p < q; p+=2)
        {
        if (gcd(p,q)>1) continue;
        sigma = 2.0*pi*p/q;
```

```
nold = 0;
for(ie = 0; ie < MAXY+2 ; ie++)
   {
     e = 8.0*ie/MAXY - 4.0 - 4.0/MAXY ;
     n = 0;
     polyold = 1.0; poly = 2.0*cos(sigma) - e;
     if( polyold*poly < 0.0 ) n++ ;

     for( m = 2; m < q/2; m++ )
        {
          polynew = (2.0*cos(sigma*m)-e)*poly-polyold;
          if( poly*polynew < 0.0) n++;
          polyold = poly; poly = polynew;
        }

     :   The corresponding program part for the even
     :   eigenfunctions has been left out here.

     if(n > nold) action;
     nold = n;
   }
 }
}
```

The **action** above is to be understood as a C command that is used, for example, to draw a point on the screen or write the corresponding pair of (σ–E) values to a data file. The function gcd(p,q) used in the third line of the program calculates the greatest common divisor of p and q.

Result

The result of this program can be seen in Fig. 2.14. The energy is plotted on the vertical axis, the parameter σ on the horizontal one. The energy is increased in steps the size of a pixel. Whenever there are eigenvalues in this small interval, a point is plotted at the corresponding position.

It is surprising and fascinating what a complex, delicate, and appealing structure is exhibited by this *Hofstadter butterfly*, which arises from the simple discrete Schrödinger equation. First of all, the symmetries $E \to -E$ and $\sigma \to 1 - \sigma$ catch the eye. Furthermore, the butterfly looks self-similar: parts of the picture reappear in different places in a slightly distorted shape, a pattern which seems to continue to the infinitely small. For each value of $\sigma = p/q$ there are q eigenvalues, some of which may be degenerate. These values, however, are distributed evenly only along the left and right edges. Otherwise they are concentrated in just a few regions. At $\sigma = 1/2$ there are just two points, which for general k and ν will turn into exactly two bands. Close to $\sigma = 1/3$ there are three bands, four in the vicinity of $\sigma = 1/4$, five around $\sigma = 1/5$ and $\sigma = 2/5$, ten at $\sigma = 1/10$ and $3/10$, and so on. Since close to any rational number $\sigma = m/n$ (m and n relatively prime) with a small

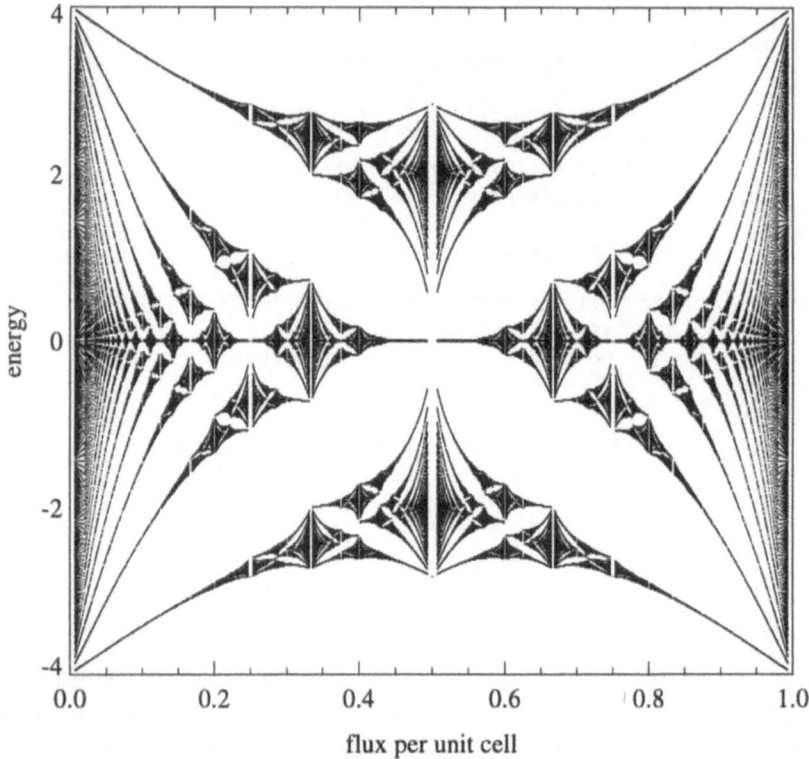

Fig. 2.14. Eigenvalues of the Harper equation as a function of $\sigma = Ba^2/(hc/e)$, i.e., of the magnetic flux through the unit cell in units of the flux quantum

denominator n there are also others with a large denominator, there will always be a large number of bands next to those values of σ that correspond to a small denominator. Any irrational number can be approximated by a sequence of rational numbers with increasing denominators. Consequently, there will always be an infinite number of energy bands next to an irrational number. This results in the artistic drawing which, given infinite resolution, resists our imagination.

Exercise

Write a program which draws a Hofstadter butterfly in a window of 500×500 pixels as quickly as possible. Take advantage of the symmetries of the butterfly and conceive an algorithm which bridges the many large holes.

Literature

Hofstadter D.R. (1976) Energy Levels and Wave Functions of Bloch Electrons in Rational and Irrational Magnetic Fields. Phys. Rev. B 14:2239

Stoer J., Bulirsch R. (1996) Introduction to Numerical Analysis. Springer, Berlin, Heidelberg, New York

2.5 The Hubbard Model

Quantum mechanics certainly is not one of the easier fields of physics. Position and momentum, wave and particle lose their intuitive meaning. The results of individual measurements generally tell us little, since quantum mechanics is, at its core, a statistical theory whose predictions are expressed as probabilistic statements. Add to this a considerable mathematical apparatus in which the concept of states has to be introduced and operators have to be explained.

The mathematical description becomes even more difficult if one wants to investigate the quantum mechanics of interacting particles, e.g., superconductivity due to interacting electrons in a solid. Since the individual electrons are in principle indistinguishable, one has to suitably construct the multiparticle state by a superposition of products of single-particle states. For electrons, this leads to the Pauli exclusion principle: Each single-particle state may only be occupied once. In this space of the multiparticle wave functions, the Hamiltonian can be represented as a matrix whose eigenvalues specify the energies of the stationary states.

There are only a few cases for which this matrix can be diagonalized by analytic methods. But even a numerical solution is not easy to obtain. The size of the system and the increase in the number of degrees of freedom turn out to be the most difficult problem here.

As an example of a quantum-mechanical multiparticle system, we consider an electron gas in a crystal lattice. In dealing with this problem, we have to take into account both the repulsive Coulomb interaction between the electrons and their interaction with the ion cores. A simplified description of this situation, which still contains all the essential elements, was suggested in 1963 by J. Hubbard. Presently this model is being discussed in connection with high-temperature superconductivity. It is, however, still uncertain whether this model exhibits superconductivity at all.

We want to demonstrate here how one can generate and diagonalize the Hamiltonian matrix of the one-dimensional Hubbard model on the computer. In doing this, however, we have to limit ourselves to a few particles.

Physics

In principle, the multiparticle states are constructed from products of single-particle states. There is, however, a formalism which allows us to get by without the explicit specification of the product wave functions; this is the so-called *second quantization* (a somewhat misleading term). It uses creation and annihilation operators, and the states can be generated by applying operators to the vacuum. In the process, the fermionic character is taken into account via the operator algebra. The Hamiltonian is represented by creation and annihilation operators as well, so that the Hamiltonian matrix can eventually be constructed algebraically.

First one has to label the single-particle states and define their order. Let $|0\rangle$ be the vacuum. A multiparticle state with N particles then has the form

$$c_{k_1}^\dagger c_{k_2}^\dagger \dots c_{k_N}^\dagger |0\rangle \ , \tag{2.43}$$

where $(k_1, k_2, \dots, k_N)$ is an ordered N-tuple of indices and k_i is the index of the corresponding single-particle wave function. This state can be represented as a determinant of single-particle wave functions $\varphi_k(r)$. For example, for two particles we have

$$\left\langle r_1, r_2 \left| c_1^\dagger c_2^\dagger \right| 0 \right\rangle = \psi(r_1, r_2)$$
$$= \frac{1}{\sqrt{2}} \left[\varphi_1(r_1) \varphi_2(r_2) - \varphi_2(r_1) \varphi_1(r_2) \right] \ . \tag{2.44}$$

As mentioned above, this explicit representation is not even necessary. If the single-particle states are orthonormal, the only important thing is the antisymmetry of the fermionic states, which is represented by the following operator algebra:

$$c_k^\dagger c_m + c_m c_k^\dagger = \delta_{km} \ ,$$
$$c_k^\dagger c_m^\dagger + c_m^\dagger c_k^\dagger = 0 \ , \tag{2.45}$$
$$c_k c_m + c_m c_k = 0 \ .$$

The operator $c_k^\dagger$ generates a single-particle wave function $\varphi_k(r)$ within the multiparticle state if that wave function is not present yet; otherwise, the result is 0. The adjoint operator c_k annihilates $\varphi_k(r)$ if present; otherwise, the result is 0 again. The Pauli exclusion principle follows from $c_k^\dagger c_k^\dagger = 0$, which implies that double occupancy of $\varphi_k(r)$ is not permitted.

In the Hubbard model, the single-particle states $\varphi_k(r)$ are wave functions of electrons localized at lattice sites. The index k contains the number of the site, and the quantum number for the z-component of the spin, represented by $\sigma = \uparrow$ or $\sigma = \downarrow$. The kinetic energy is taken into consideration via the "hopping" of electrons from a site k to a neighboring site m, described by the operator

$$- t \ c_{m\sigma}^\dagger c_{k\sigma} \ . \tag{2.46}$$

If two electrons occupy the same site k, then they are supposed to feel the Coulomb repulsion

$$U n_{k\uparrow} n_{k\downarrow} \ . \tag{2.47}$$

Here, $n_{k\sigma} \equiv c_{k\sigma}^\dagger c_{k\sigma}$ yields the value 1, if an electron with the spin σ is located at the site k; otherwise, $n_{k\sigma}$ yields the value 0. The parameters t and U model the magnitudes of the kinetic and potential energy respectively.

In the following, we limit ourselves to a chain with M sites and assume periodic boundary conditions. Then the Hamiltonian is

$$H = -t \sum_{k=1}^{M} \sum_{\sigma} \left(c_{k\sigma}^{\dagger} c_{k+1,\sigma} + c_{k+1,\sigma}^{\dagger} c_{k\sigma} \right) + U \sum_{k=1}^{M} n_{k\uparrow} n_{k\downarrow} \qquad (2.48)$$

with $c_{M+1,\sigma}^{\dagger} = c_{1,\sigma}^{\dagger}$ and $c_{M+1,\sigma} = c_{1\sigma}$. For the order of the single-particle states, we choose

$$\{1\uparrow, 2\uparrow, \ldots, M\uparrow\}, \{1\downarrow, 2\downarrow, \ldots, M\downarrow\} . \qquad (2.49)$$

Thus every multiparticle state can be described by the occupancy numbers of the single-particle states. Then we get, for example:

$$|\{1,1,0,\ldots,0\}, \{0,\ldots,0,1\}\rangle = c_{1\uparrow}^{\dagger} c_{2\uparrow}^{\dagger} c_{M\downarrow}^{\dagger} |0\rangle .$$

If one applies the operator $c_{k\sigma}^{\dagger}$ to a state, then one has to swap this operator with the other operators $c_{m\sigma'}^{\dagger}$, using the algebra (2.45), until the sequence is in the correct order (2.49). By this procedure one also obtains the sign of the state. For example, we get

$$\begin{aligned}
c_{1\downarrow}^{\dagger} |\{1,0,\ldots,0\}, \{0,\ldots,0,1\}\rangle &= c_{1\downarrow}^{\dagger} c_{1\uparrow}^{\dagger} c_{M\downarrow}^{\dagger} |0\rangle \\
&= -c_{1\uparrow}^{\dagger} c_{1\downarrow}^{\dagger} c_{M\downarrow}^{\dagger} |0\rangle \\
&= -|\{1,0,\ldots,0\}, \{1,0,\ldots,0,1\}\rangle
\end{aligned}$$

and

$$\begin{aligned}
c_{M\downarrow} |\{1,0,\ldots,0\}, \{0,\ldots,0,1\}\rangle &= c_{M\downarrow} c_{1\uparrow}^{\dagger} c_{M\downarrow}^{\dagger} |0\rangle \\
&= -c_{1\uparrow}^{\dagger} c_{M\downarrow} c_{M\downarrow}^{\dagger} |0\rangle \\
&= -c_{1\uparrow}^{\dagger} |0\rangle + c_{1\uparrow}^{\dagger} c_{M\downarrow}^{\dagger} c_{M\downarrow} |0\rangle \\
&= -c_{1\uparrow}^{\dagger} |0\rangle \\
&= -|\{1,0,\ldots,0\}, \{0,\ldots,0\}\rangle .
\end{aligned}$$

Generally, when applying $c_{k\sigma}^{\dagger}$ or $c_{k\sigma}$ to a state

$$|n\rangle = |\{n_{1\uparrow}, \ldots, n_{M\uparrow}\}, \{n_{1\downarrow}, \ldots, n_{M\downarrow}\}\rangle ,$$

the number of particles to the left of $k\sigma$ determines the sign. Therefore we define the sign function:

$$\mathrm{sign}(k\sigma, n) = (-1)^{\delta_{\sigma\downarrow} \sum_{i=1}^{M} n_{i\uparrow}} (-1)^{\sum_{j=1}^{k-1} n_{j\sigma}} . \qquad (2.50)$$

The function produces as many factors -1 as there are non-zero entries in n in front of the position $k\sigma$. This fact allows us to write the effect of the creation and annihilation operators in the following manner:

$$c_{k\sigma}^{\dagger} |n\rangle = (1 - n_{k\sigma}) \mathrm{sign}(k\sigma, n) |\{n_{1\uparrow}, \ldots, 1, \ldots, n_{M\downarrow}\}\rangle . \qquad (2.51)$$

The number 1 is at the position $k\sigma$, and the factor $(1 - n_{k\sigma})$ ensures that there is no double occupancy of the same state. Correspondingly we get

$$c_{k\sigma} |n\rangle = n_{k\sigma} \text{sign}(k\sigma, n) |\{n_{1\uparrow}, \ldots, 0, \ldots, n_{M\downarrow}\}\rangle , \qquad (2.52)$$

which annihilates the state $k\sigma$.

Every site of the chain has four possible states – vacant, singly occupied by either $\uparrow$ or $\downarrow$, or doubly occupied with both $\uparrow$ and $\downarrow$ – so there are altogether 4^M multiparticle states. In this basis the Hamiltonian H is therefore a $4^M \times 4^M$-matrix. But first of all this matrix is mostly filled with zeros, and second it can be divided into submatrices by taking advantage of symmetries. The following quantities are conserved in the Hubbard model: the number of particles with positive $(N_\uparrow)$ and negative $(N_\downarrow)$ spins, the quantum number of the total spin, and spatial symmetries. This greatly reduces the size of these submatrices of H. For three sites, for example, the 64×64 matrix can be reduced to submatrices with a maximum size of 3×3.

Here, we only want to consider $N_\uparrow$ and $N_\downarrow$ as conserved quantities. Also, we do not want to make a secret of the fact that a number of exact results, which can be obtained from methods using a Bethe ansatz, are known for the one-dimensional model. Thus our example only serves as a simplified introduction to numerical calculations for two- or three-dimensional models.

Algorithm

Once again, we want to use *Mathematica* to concisely formulate the representation of the operator algebra (2.45) by (2.51) and (2.52). Instead of the symbol $|\ldots\rangle$, we use the header s in *Mathematica* and write the multiparticle states in the form

$$\texttt{s[arg]}, \quad \text{where } \texttt{arg} = \{\{n_{1\uparrow}, \ldots, n_{M\uparrow}\}, \{n_{1\downarrow}, \ldots, n_{M\downarrow}\}\} \qquad (2.53)$$

and $n_{k\sigma} \in \{0, 1\}$. We need the header s because we have to add states and multiply them by scalars. If we were to perform these operations on the list arg itself, we would get incorrect results. The main manipulations, however, concern the argument of s. For example, one obtains $n_{k\sigma}$ from arg[[sigma, k]] with k $= k$, sigma $= 1$ and sigma $= 2$ respectively, for $\sigma = \uparrow$ and $\sigma = \downarrow$.

After specifying the numbers M, $N_\uparrow$, and $N_\downarrow$ we use the functions Permutations[...] and Table[...] to generate the list index, which contains all possible states. The command Flatten[...] removes one level of braces. For $M = 3$, $N_\uparrow = 2$, and $N_\downarrow = 1$, for example, we obtain

```
index = {{{1,1,0}, {1,0,0}}, {{1,1,0}, {0,1,0}},

        {{1,1,0}, {0,0,1}}, {{1,0,1}, {1,0,0}},

        {{1,0,1}, {0,1,0}}, {{1,0,1}, {0,0,1}},
```

{{0,1,1}, {1,0,0}}, {{0,1,1}, {0,1,0}},

{{0,1,1}, {0,0,1}}}}

The operator $c_{k\sigma}^\dagger$ has to generate a 1 in the right spot in the argument of s. This is done by the function plus:

```
plus[k_,sigma_][arg_]:= ReplacePart[arg,1,{sigma,k}]
```

Correspondingly, minus generates a 0. Using the sign function (2.50) introduced above, we can now define the operator $c_{k\sigma}^\dagger$ by its effect on states, analogously to (2.51) and (2.52):

```
cdagger[k_,sigma_][factor_.*s[arg_]]:= factor*
                        (1 - arg[[sigma,k]])*
                        sign[k,sigma,arg]*
                        s[plus[k,sigma][arg]]
```

The parameter factor_. is set to the value 1 in *Mathematica*, if the state does not have an additional factor. Since these operators can always generate the value 0 – e.g., $c_{1\uparrow}^\dagger|\{1,\ldots\},\{\ldots\}\rangle = 0$ – cdagger has to be defined for the number 0 as well:

```
cdagger[k_,sigma_][0]:= 0
```

The operator $c_{k\sigma}$ is defined by c[k_,sigma_] in an analogous way. In addition we still need the operators $n_{k\sigma}$:

```
            n[k_,sigma_][0]:= 0
n[k_,sigma_][factor_.*s[arg_]]:= factor*
                        arg[[sigma,k]]*s[arg]
```

Then the Hamiltonian (2.48) of the Hubbard chain can simply be written as

```
H[vector_]=Expand[
            -t*Sum[cdagger[k,sigma][c[k+1,sigma][vector]]+
                cdagger[k+1,sigma][c[k,sigma][vector]],
                {k,sites},{sigma,2} ]+
            u*Sum[n[k,1][n[k,2][vector]]  ,{k,sites}]
            ]
```

Here, the length M of the chain is designated by sites. In order to obtain the Hamiltonian matrix, we need the scalar products $\langle n_i|H|n_j\rangle$ of $|n_i\rangle$ with $H|n_j\rangle$. Since our multiparticle states $|n\rangle$ are orthonormal, we only have to define the linearity of the scalar product:

```
scalarproduct[a_,0]:= 0
scalarproduct[a_,b_ + c_]:= scalarproduct[a,b] +
                        scalarproduct[a,c]
scalarproduct[s[arg1_],factor_. s[arg2_]]:=
                        factor*If[arg1==arg2,1,0]
```

The antilinearity in the first argument is not needed here. Note that the comparison of two states, which, in this case, amounts to the comparison of two nested lists, can be written in a very compact manner in *Mathematica* by using `arg1==arg2`.

The following table yields the desired Hamiltonian matrix $\langle n_i|H|n_j\rangle$:

```
h = (hlist = Table[H[s[index[[j]]]], {j, end}];
      Table[ scalarproduct[ s[index[[i]]], hlist[[j]]],
                            {i,end}, {j,end}])
```

The upper limit for the loop indices has been determined beforehand via `end = Length[index]`.

With `Eigenvalues[...]` and `Eigensystem[...]`, one then obtains the energies and the stationary states of the multiparticle problem. We filter out the ground state with the command sequence

```
g[uu_]:= Sort[Thread[ Eigensystem[
            N[ h /. {t -> 1.0, u -> uu} ]]]][[1,2]]
```

The command `Thread[...]` arranges the result of `Eigensystem[...]` in such a way that each eigenvalue is combined with its associated eigenvector. Then the ground state, being the state with the smallest energy, can be found in the first position of the sorted result. Its first component is the ground state energy; the second component, a list, contains the coefficients g_n of the ground state $|g\rangle$ with respect to the basis $|n\rangle$). We are interested, for example, in the fraction of sites with double occupancy in the ground state, i.e., in

$$\frac{1}{M}\sum_{k=1}^{M}\langle g\,|n_{k\uparrow}n_{k\downarrow}|\,g\rangle = \frac{1}{M}\sum_{k=1}^{M}\sum_{\{n\}}|g_n|^2\,n_{k\uparrow}(n)\,n_{k\downarrow}(n)\ . \tag{2.54}$$

Mathematica's version of the right-hand side of (2.54) is

```
Sum[Sum[Abs[g[u][[j]]]^2*index[[j,1,k]]*index[[j,2,k]],
         {j,end}], {k,sites}] / sites
```

Since the sums are interchangeable and can be regarded as scalar products, this can be formulated in the following very concise manner:

```
(Abs[g[u]]^2).Map[#[[1]].#[[2]]&, index] / sites
```

Results

For $M = 3$ sites and $N_\uparrow = 2$, $N_\downarrow = 1$, *Mathematica* is still capable of finding the energies analytically as a function of U and t, thanks to the symmetry of the matrix. To see this, we write the result to a file in TeX format, using

```
TeXForm[Eigenvalues[h]] >> hubbard.tex
```

add the mathematics environment and a few formatting commands, and compile with TEX or LATEX. The result is a list of the eigenvalues of the 9×9 Hubbard matrix:

$$\left\{ 0, u, u, \frac{2u}{3} - \frac{2^{\frac{1}{3}}\left(-27t^2-u^2\right)}{3f(u,t)^{\frac{1}{3}}} + \frac{f(u,t)^{\frac{1}{3}}}{3\cdot 2^{\frac{1}{3}}}, \frac{2u}{3} - \frac{2^{\frac{1}{3}}\left(-27t^2-u^2\right)}{3f(u,t)^{\frac{1}{3}}} + \frac{f(u,t)^{\frac{1}{3}}}{3\cdot 2^{\frac{1}{3}}}, \right.$$

$$\frac{2u}{3} + \frac{\left(1+i\sqrt{3}\right)\left(-27t^2-u^2\right)}{3\cdot 2^{\frac{2}{3}}f(u,t)^{\frac{1}{3}}} - \frac{\left(1-i\sqrt{3}\right)f(u,t)^{\frac{1}{3}}}{6\cdot 2^{\frac{1}{3}}},$$

$$\frac{2u}{3} + \frac{\left(1+i\sqrt{3}\right)\left(-27t^2-u^2\right)}{3\cdot 2^{\frac{2}{3}}f(u,t)^{\frac{1}{3}}} - \frac{\left(1-i\sqrt{3}\right)f(u,t)^{\frac{1}{3}}}{6\cdot 2^{\frac{1}{3}}},$$

$$\frac{2u}{3} + \frac{\left(1-i\sqrt{3}\right)\left(-27t^2-u^2\right)}{3\cdot 2^{\frac{2}{3}}f(u,t)^{\frac{1}{3}}} - \frac{\left(1+i\sqrt{3}\right)f(u,t)^{\frac{1}{3}}}{6\cdot 2^{\frac{1}{3}}},$$

$$\left. \frac{2u}{3} + \frac{\left(1-i\sqrt{3}\right)\left(-27t^2-u^2\right)}{3\cdot 2^{\frac{2}{3}}f(u,t)^{\frac{1}{3}}} - \frac{\left(1+i\sqrt{3}\right)f(u,t)^{\frac{1}{3}}}{6\cdot 2^{\frac{1}{3}}} \right\},$$

where we have manually added the abbreviation

$$f(u,t) = -162t^2u + 16u^3 - 18u\left(-9t^2 + u^2\right)$$
$$+ \sqrt{4\left(-27t^2 - u^2\right)^3 + \left(-162t^2u + 16u^3 - 18u\left(-9t^2 + u^2\right)\right)^2}$$

for clarity.

Figure 2.15 shows these energies as a function of the ratio U/t between Coulomb energy and kinetic energy. For $U = 0$, there is no interaction between the electrons, so we can calculate the energy levels simply by filling the single-particle levels. In the case of periodic boundary conditions, these single-particle energies can be specified explicitly for any M. We define the wave vectors $k_\nu = 2\pi\nu/M$, $\nu = 0, 1, \ldots, M - 1$. Then the single-particle eigenstates of the Hubbard Hamiltonian are given by

$$|k_\nu^\sigma\rangle = \frac{1}{\sqrt{M}} \sum_{l=1}^{M} \exp\left(-ik_\nu l\right) c_{l\sigma}^\dagger |0\rangle \ ,$$

$$E_\nu = -2t \cos k_\nu \ .$$

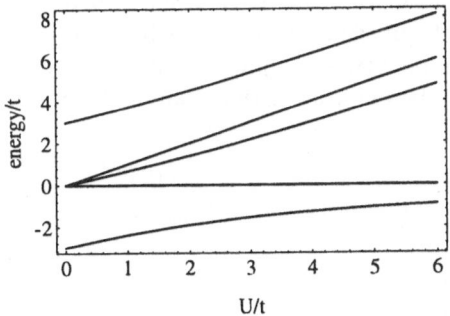

Fig. 2.15. Energy eigenvalues of the Hubbard model as a function of U/t for $M = 3$, $N_\uparrow = 2$, and $N_\downarrow = 1$

For $M = 3$, we obtain the single-particle energies $-2t$, t, and t again. These levels can be occupied by two spin ↑ particles and one spin ↓ particle with the restriction that double occupancy by two ↑-spins is forbidden. This results in the following nine levels, in agreement with Fig. 2.15:

twofold degenerate: $\varepsilon_1 = -3t$,
fivefold degenerate: $\varepsilon_2 = 0$,
twofold degenerate: $\varepsilon_3 = 3t$.

The fact that, even for $U \neq 0$, all levels but the one with $\varepsilon = 0$ are twofold degenerate is due to the translational and mirror symmetry of this model. The average double occupancy rate in the ground state for $M = 4$ sites and the so-called *half-filled case*, i.e., $N_\uparrow + N_\downarrow = M$ (in this specific case for $N_\uparrow = 2$ and $N_\downarrow = 2$) is shown in Fig. 2.16. For $U \to \infty$ double occupancy is obviously forbidden. Figures 2.17 and 2.18 show results for a Hubbard ring with $M = 6$ sites, $N_\uparrow = 3$, and $N_\downarrow = 3$. As moderate as these numbers may seem, the number of states strongly increases with M, and for this example it is already $\binom{6}{3} \cdot \binom{6}{3} = 400$. Calculating the energies of the stationary states thus means determining the eigenvalues of a 400×400 matrix. If not for the symmetry and the multiple degeneracy, the diagram in Fig. 2.17 would show 400 different energy levels. For the case of a half-filled band, i.e., $(N_\uparrow + N_\downarrow)/M = 1$, the ground state energy per site can be specified analytically in the limit $M \to \infty$, for arbitrary values of U/t, if the additional condition $N_\uparrow/M = N_\downarrow/M = 1/2$ is fulfilled. One obtains

$$\frac{\varepsilon_0}{M} \longrightarrow -4t \int_0^\infty \frac{J_0(\omega) J_1(\omega) \, d\omega}{\omega \left[1 + \exp(\omega U / (2t))\right]} \quad \text{as } M \to \infty , \tag{2.55}$$

where J_0 and J_1 are Bessel functions. We evaluate the integral numerically and compare the result to the corresponding values for $M = 6$. Surprisingly, the two curves are not very far apart at all, as can be seen in Fig. 2.18.

In order to calculate larger systems, one must do two things: first, take the symmetry into account and second, use a fast programming language. Thus it is currently possible to obtain an exact diagonalization for models with up to $M = 25$ sites. Using the Quantum Monte Carlo method, which treats the Hubbard model in an exact numerical way (except for statistical errors) one can investigate up to $M = 1024$ lattice sites. Thus, even with today's supercomputers only the smallest of quantum systems can be treated. A dramatic increase in the capabilities of future computers will only allow a minor increase of the system size – unless new algorithms can be developed.

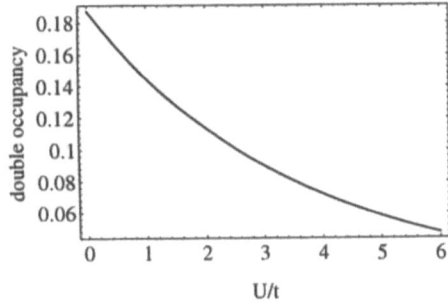

Fig. 2.16. Average double occupancy in the ground state as a function of U/t for $M = 4$, $N_\uparrow = 2$, and $N_\downarrow = 2$

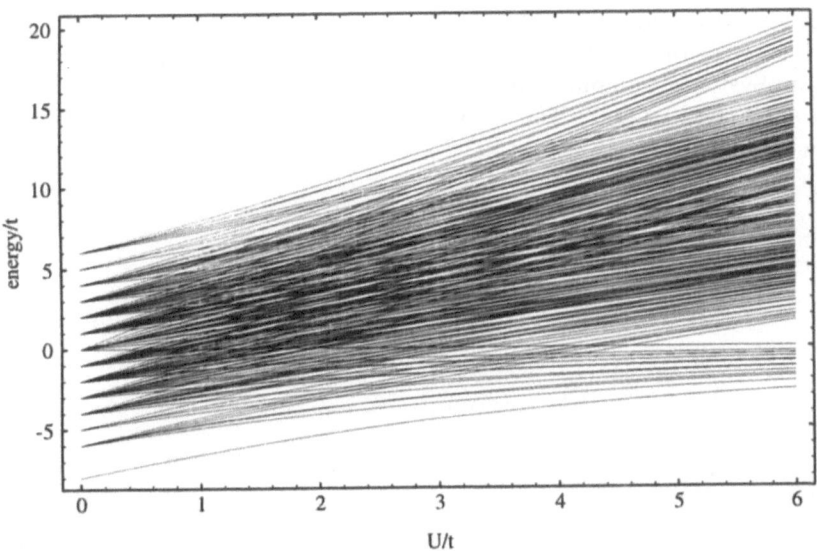

Fig. 2.17. Energy eigenvalues of the Hubbard model as a function of U/t for $M = 6$, $N_\uparrow = 3$, and $N_\downarrow = 3$

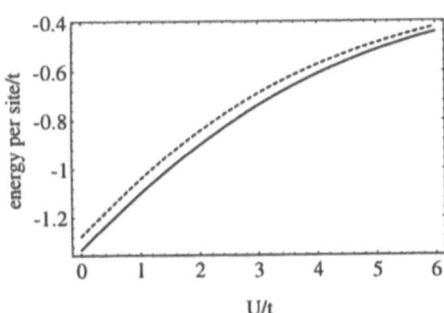

Fig. 2.18. Ground state energy per site for the half-filled band. The dashed curve is the exact result for $M \to \infty$, the solid curve gives the values for $M = 6$

Exercise

The z-component of the spin operator at the site i can be expressed by the particle number operators $n_{i\uparrow}$ and $n_{i\downarrow}$: $S_i^z = (n_{i\uparrow} - n_{i\downarrow})/2$. For the ground state $|g\rangle$ of the Hubbard model with $M = 4$ sites and $N_\uparrow = N_\downarrow = 2$, calculate the spin correlations $(1/M) \sum_{i=1}^M \langle g|S_i^z S_{i+1}^z|g\rangle$ and $(1/M) \sum_{i=1}^M \langle g|S_i^z S_{i+2}^z|g\rangle$ as a function of U/t.

Literature

Hirsch J.E. (1985) Two-dimensional Hubbard Model: Numerical Simulation Study. Phys. Rev. B 31:4403

Lin H.Q., Gubernatis J.E. (1993) Exact Diagonalization Methods for Quantum Systems. Computers in Physics 7:400

Montorsi A. (Ed.) (1992) The Hubbard Model: A Reprint Volume. World Scientific, Singapore

3. Iterations

A function consists of a set of instructions that convert given input values to output values. Now if the output itself is part of the domain of the function considered, it can become an input in turn, and the function returns new output values. This process can then be repeated indefinitely. While there are often no analytic methods to calculate the structural properties of such iterations of the form $x_{t+1} = f(x_t)$, they are easily generated on the computer. Obviously, one just has to apply the same function $f(x)$ to the result repeatedly. We want to demonstrate this with a few examples.

3.1 Population Dynamics

Arguably, the best-known iteration of a nonlinear function is the so-called *logistic map*. It is a simple parabola that projects the real numbers of the unit interval onto themselves. We want to present it here because every scientist should know the properties of such an elementary iteration.

A simple mechanism leads to complex behavior – this is the result of the repeated application of a quadratic function. Although there are only a few analytic results on this, anyone can easily reproduce the iteration with a pocket calculator. For certain parameter values of the parabola the iterated numbers jump around in the unit interval in an irregular manner. The sequence of numbers is extremely sensitive to changes of the starting value of the iteration. A well-defined function definition yields a sequence of numbers that is practically unpredictable even for a small initial uncertainty. Such a behavior is called *deterministic chaos*.

Some quantities that can be calculated with the computer are not just the result of a mathematical folly; rather, one finds these numbers in many experiments near the transition from regular to chaotic behavior. Such a universality of quantitative values is an additional fascinating aspect of this equation.

Physics

We consider the iteration

$$x_{n+1} = 4rx_n(1 - x_n) = f(x_n) , \quad n = 0, 1, 2, 3, \dots , \tag{3.1}$$

which was first introduced in 1845 by the Belgian biomathematician P.F. Verhulst, in a work on population dynamics. Here, x_n is iterated in the interval $[0, 1]$, and r is a parameter that can be varied in the range $[0, 1]$ as well.

The function (3.1) is of interest for various questions:

1. As a simple model for the time development of a growing population. In this model, x_n is proportional to the population density of a species at time n. Its growth is described by a linear contribution $4rx_n$ which, taken by itself, leads to exponential growth, i.e., a population explosion, for $r > 1/4$. The population growth is limited, however, by the nonlinear contribution $-4rx_n^2$ (due, e.g., to a limited food supply).

2. As a simple example of classical mechanics with friction. We will see in Sect. 4.2 that the motion of a mechanical system can be analyzed by a Poincaré section in phase space. This map results in an iteration of a nonlinear function. Therefore, it is useful to know what properties such iterations already have in one dimension, even though one-dimensional Poincaré sections in mechanics cannot exhibit chaos.

3. As an example of a (bad) numerical solution of a differential equation. If neighboring x_n values differ only by small amounts, then n and x_n can be extended to real numbers in a continuous way. If we set

$$x_{n+1} - x_n \simeq \frac{dx}{dn} ,$$

then (3.1) becomes an approximation of the differential equation

$$\frac{dx}{dn} = (4r - 1)x - 4rx^2$$

with the solution

$$x(n) = \frac{4r - 1}{4r + \text{const} \cdot \exp\left((1 - 4r)\,n\right)} .$$

For $r < 1/4$ this equation has the attractor $x_\infty = 0$, whereas for $r > 1/4$ all initial values $x(0) > 0$ lead to the value $x_\infty = 1 - 1/(4r)$. Like the discrete iteration, this equation has a phase transition at $r_0 = 1/4$ from a vanishing population to a constant one. It will turn out, however, that for larger values of r, the differential equation no longer bears any resemblance to its discrete approximation.

For $r > r_0 = 1/4$, (3.1) has a fixed point $x^* = f(x^*)$ with $x^* = 1 - 1/(4r)$. We now consider a small deviation ε_0 from the fixed point and iterate f for $x^* + \varepsilon_0$. Because of the relation $f(x^* + \varepsilon_0) \simeq f(x^*) + f'(x^*)\varepsilon_0 = x^* + f'(x^*)\varepsilon_0$ we have for small ε_n:

$$\varepsilon_n \simeq f'(x^*)\varepsilon_{n-1} \simeq [f'(x^*)]^n \varepsilon_0 . \tag{3.2}$$

Therefore, the perturbation ε_0 explodes for $|f'(x^*)| > 1$, which corresponds to $r > r_1 = 3/4$. We will see that, in this case, the x_n jump back and forth between the values x_1^* and x_2^* after a few steps:

$$x_1^* = f(x_2^*) , \ x_2^* = f(x_1^*) ,$$
$$\text{so } x_i^* = f(f(x_i^*)) \text{ for } i = 1, 2 . \tag{3.3}$$

This means that for certain values of r, the iterated function $f^{(2)}(x) = f(f(x))$ has two stable fixed points.

Now $|f^{(2)\prime}(x_i^*)| > 1$ for $r > r_2 = (1+\sqrt{6})/4$. Therefore, the same argument as above shows that the cycle of length 2 (or two-cycle, for short) becomes unstable for $r > r_2$; the x_n run into a four-cycle. This doubling of periods continues until their number reaches infinity at $r_\infty \simeq 0.89$.

A doubling of the period means reducing the frequency by half. Consequently, subharmonics in the frequency spectrum are generated. At r_l, the 2^{l-1}-cycle becomes unstable, resulting in a 2^l-cycle. As $l \to \infty$, the r_l values get closer and closer to r_∞, according to the following pattern:

$$r_l \approx r_\infty - \frac{c}{\delta^l} . \tag{3.4}$$

From this we can conclude

$$\delta = \lim_{l \to \infty} \frac{r_l - r_{l-1}}{r_{l+1} - r_l} . \tag{3.5}$$

The number δ is called the *Feigenbaum constant* and has the value $\delta = 4.6692\ldots$. It receives its importance from the fact that it is a universal quantity, i.e., it is observed for many different functions $f(x)$. Even experiments with real nonlinear systems that undergo a transition to chaos via period doubling obtain the same number.

For larger values of r, one obtains chaotic bands: x_n appears to jump around randomly, without a period, in one or more intervals. This holds true even though the x_n values are not generated randomly, but by the well-defined parabola $f(x)$ from (3.1). As mentioned before, such a behavior is called deterministic chaos.

In the chaotic region, one can use the simple parabola to study an essential property of deterministic chaos, the sensitivity to the initial conditions. Two extremely close values separate after just a few iterations and jump around in the chaotic bands seemingly independently. For a quantitative characterization, we designate the two orbits by x_n and $x_n + \varepsilon_n$. Then, for $\varepsilon_0 \to 0$

$$\varepsilon_n \simeq \varepsilon_0 e^{\lambda n} . \tag{3.6}$$

The parameter λ is called the Lyapunov exponent; it is a measure of the sensitivity to the initial state. A positive exponent $\lambda > 0$ indicates chaos. We have

$$\left| f^{(n)}(x_0 + \varepsilon_0) - f^{(n)}(x_0) \right| \simeq \varepsilon_0 e^{\lambda n} . \tag{3.7}$$

In the limit $\varepsilon_0 \to 0$ this yields the derivative of the n-fold iterated function $f^{(n)}(x_0) = f(f(\cdots f(x_0))\cdots)$. With the compound rule of three one immediately gets

$$\lambda = \lim_{n \to \infty} \frac{1}{n} \sum_{i=0}^{n-1} \ln |f'(x_i)| \;. \tag{3.8}$$

Thus, in order to determine λ, only one orbit $\{x_i\}$ needs to be calculated and the logarithms of the slope of f at the points x_i have to be added up.

Algorithm

With *Mathematica* one can obtain an n-fold iteration of a given function, such as f[x_] = 4 r x (1-x), simply by using

    iterf[n_]:= Nest[f,x,n]

The determination of the bifurcation points r_l and consequently the Feigenbaum constant δ is significantly more difficult, however. One can find r_1 and r_2 by solving the two equations, $f^{(l)}(x^*) = x^*$ and $|df^{(l)}(x^*)/dx| = 1$, for the unknown variables x^* and r_l, either by hand or by using the *Mathematica* function FindRoot. But we were unable to determine higher r_l values this way. It is sufficient, however, to consider so-called superstable periodic orbits. These are orbits that start at $x_0 = 1/2$ and return there after 2^l steps. Because $f'(1/2) = 0$, deviations from x_0 converge towards the 2^l-cycle very quickly. In each interval $[r_l, r_{l+1}]$ there is one superstable orbit with a period 2^l, at a value R_l. Here, the R_l values, like the r_l values, scale with the Feigenbaum constant δ as they approach r_∞ (see (3.5)).

Now again, solving the equation

$$f^{(2^l)}\left(\frac{1}{2}\right) = \frac{1}{2}$$

numerically is not easy. Since this equation has $2^l - 1$ additional fixed points, the function oscillates very rapidly and FindRoot fails. But there is a trick: one inverts the equation above. Since the inverse of f has two branches, one has to specify in the iteration whether x_n lies to the right (R) or left (L) of $x_0 = 1/2$. Therefore, any periodic orbit is determined by a sequence of C, R and L, where C (=center) denotes the initial value $x_0 = 1/2$ and R, L indicate whether x_i lies to the right or left of the maximum. Thus, the iteration is described by *symbolic dynamics* subject to certain rules.

There is a simple algorithm for determining the sequence of Rs and Ls for a given period 2^l. We describe it without proof. The initial value is supposed to be $x_0 = 1/2$, i.e., the sequence always starts with a C. For $l = 1$, we obviously get CR. The sequence for $l + 1$ is constructed from the one for l by doubling the l-sequence and replacing the C in the middle by L if the number of Rs in the l-sequence is odd, or by R if it is even. For $l = 2$, the doubling yields $CRCR$ and this becomes $CRLR$, as CR contains just one R. For $l = 3$, one gets $CRLRRRLR$.

The inverse iteration is most easily performed in a coordinate system whose origin is at $(x, y) = (1/2, 1/2)$ and whose two axes are scaled by a common factor such that the peak of the parabola in the new system is at $(0, 1)$. In this coordinate system, the original parabola $f(x) = 4rx(1 - x)$ takes the form

$$g(\xi) = 1 - \mu\xi^2 ,$$

where μ and r are related via

$$\mu = r(4r - 2) \quad \text{or} \quad r = \frac{1}{4}\left(1 + \sqrt{1 + 4\mu}\right) .$$

We designate the left branch of the parabola by $g_L(\xi)$ and the right one by $g_R(\xi)$. Then the superstable orbit for $l = 2$ obeys the equation

$$g_R(g_L(g_R(1))) = 0$$

or

$$1 = g_R^{-1}(g_L^{-1}(g_R^{-1}(0)))$$

with

$$g_R^{-1}(\eta) = \sqrt{\frac{1 - \eta}{\mu}} \quad \text{and} \quad g_L^{-1}(\eta) = -\sqrt{\frac{1 - \eta}{\mu}} .$$

This finally leads to the following equation for μ:

$$\mu = \sqrt{\mu + \sqrt{\mu - \sqrt{\mu}}} , \tag{3.9}$$

which can be solved by iteration or by using FindRoot. The sequence $CRLR$ determines the signs of the nested roots in (3.9). After the initial CR, which does not enter the equation above, L yields a positive and R a negative sign. This holds for all superstable periodic orbits. Thus, the orbit of period 5, with the sequence $CRLRR$, which is located between chaotic bands, is found at a μ value that obeys the following equation:

$$\mu = \sqrt{\mu + \sqrt{\mu - \sqrt{\mu - \sqrt{\mu}}}} .$$

We have programmed the generation of the sequence in *Mathematica* in the following way: For each symbol R (L), we write the numbers 1 (0); the sequence of length 2^n is stored as a list period[n]. Initially we define the list period[1]= {c,1} for $n = 1$, and the doubling of the sequence is done by

```
period[n_]:= period[n]=
    Join[period[n-1], correct[period[n-1]]]
```

Here, the symbol c still has to be replaced by 0, if the frequency with which the number 1 appears (= sum of the array elements) is odd, or by 1 if the sum is even.

```
correct[list_]:=Block[{sum=0,li=list,l=Length[list]},
                Do[sum+=li[[i]],{i,2,l}];
                If[OddQ[sum],li[[1]]=0,li[[1]]=1];
                li]
```

The multiply nested root from the equation for μ is obtained by

```
g[n_,mu_]:=Block[{x=Sqrt[mu],l=Length[period[n]]},
    Do[x=Sqrt[mu+(-1)^(period[n][[i]]) x],{i,1,3,-1}];
        x]
```

Finally, this equation is solved numerically, which requires increasing the accuracy with which FindRoot calculates the solution:

```
prec=30
maxit=30
fr[n_]:=fr[n]=(find=FindRoot[g[n,mu]==mu,
                {mu,{15/10,16/10}},
                AccuracyGoal->prec,
                WorkingPrecision->prec,
                MaxIterations->maxit];
                mu/.find)
```

In order to generate and display the orbits $x_0, x_1, x_2, x_3, \ldots$ for all parameters r (or μ), one should use a fast programming language, if possible, since each value of r requires about 1000 iterations. In principle, however, this is very simple: For the r value, one picks, for example, the x-coordinate of the pixels on the screen. For each value of r, one first calculates about 100 values of x_n, until the x_n have come very close to the attractor. Then, one calculates 1000 additional values of x_n, transforms them into the y-coordinate of the pixels, and plots a point at each pixel (x, y) obtained in this way. In C, this is written as:

```
xit = 4.*r*xit*(1.-xit);
putpixel( r*MAXX, (1.-xit)*MAXY,WHITE);
```

The parameter r is incremented or decremented one step at a time in response to keyboard input.

The density of the x_n values is of interest as well. To display it, the x-axis is subdivided into small intervals, e.g., of the order of the distance between adjacent pixels. Each interval is associated with a y-coordinate, which is initially set to the value $y = 0$. In the iteration of the parabola, for a fixed value of r, each x_n value falls into one of these intervals, whereupon the corresponding y-coordinate is incremented by one pixel and a point is plotted at the resulting (x, y) pixel. The result is a histogram of the x_n values which corresponds to the density of points on the attractor. This is done with the following function:

```
void channel( double r)
{
    double xit=.5;
    int x,y[600],i;
    clearviewport();
    for (i=0;i<MAXX;i++) y[i]=0;
    rectangle(0,0,MAXX,300);
    while(!kbhit())
    {   xit=4.*r*xit*(1.-xit);
        x=xit*MAXX;
        y[x]++;
        putpixel(x,300-y[x],WHITE);
    }
    getch();getch();clearviewport();return;
}
```

Results

Figure 3.1 shows the iteration starting from $x_0 = 0.65$ for $r = 0.87$. The x_n values move toward an attractor of period 4, which is described by the fixed points of the fourfold iterated function $f(x)$. These can be seen in Fig. 3.2. The function $f^{(4)}(x)$ intersects the straight line $y = x$ eight times, but only four of those are stable fixed points. As r is increased, attractors with the periods $8, 16, 32, \ldots$ are generated until the transition to chaos occurs at $r_\infty = 0.89 \ldots$. This is clearly visible in Figs. 3.3 and 3.4, in which 1000 x_n values are plotted for each value of r, after an initial transition period of about 100 steps. Thus, attractors with a period p are visible as p points, whereas chaotic orbits fill several intervals on the vertical. Since at most 1000 points were plotted for each value of r, one also gets an impression of the density of x_n values. This, however, is more clearly visible in the histogram in Fig. 3.5. Periodic orbits with period p show up as a series of lines with p peaks, whereas chaotic orbits jump back and forth between various bands, each with continuous densities. Figure 3.5 shows the chaotic motion just

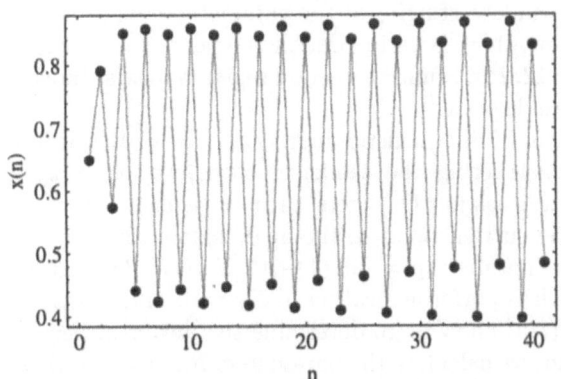

Fig. 3.1. Iteration starting from $x_0 = 0.65$ using the parameter $r = 0.87$. The x_n values move towards an attractor of period 4

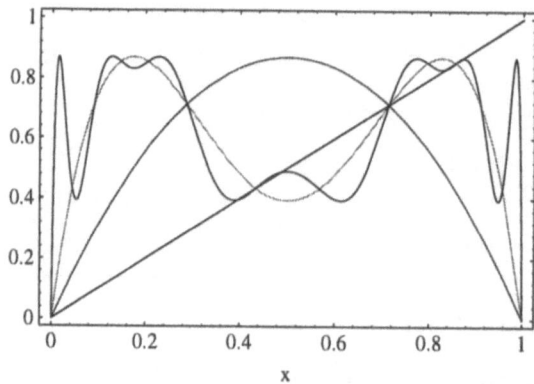

Fig. 3.2. The function $f(x)$ and its two- and fourfold iterations. The latter function has eight fixed points

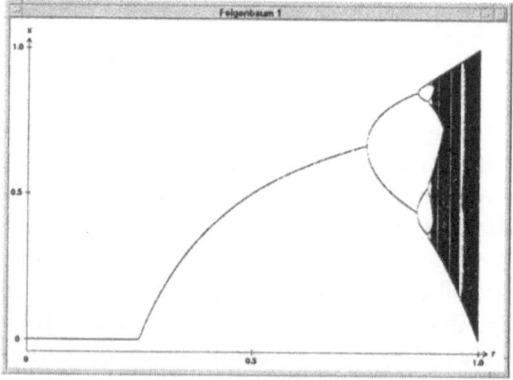

Fig. 3.3. Each vertical line contains 1000 x_n values from the iterated parabola, not showing the first 100 points. The parameter r varies between 0 and 1 along the horizontal

below a five-cycle; though there is just one band there, one still sees the five peaks of the adjacent cycle. It can be shown that only one band with density

$$\rho(x) = \frac{1}{\pi\sqrt{x(1-x)}}$$

exists for $r = 1$. In Fig. 3.4, in the chaotic region, one sees many windows with periodic orbits. The largest window, starting at the parameter $r = (\sqrt{8}+1)/4$ has a period 3. In this window, there is period doubling again: as r increases, one obtains the periods $3, 6, 12, 24, \ldots$ until a chaotic region appears. Indeed, it can be shown that all periods occur.

An additional impression of the logistic map $x_{n+1} = f(x_n)$ is obtained by geometrically constructing the sequence of points $(x_n, f(x_n))$, $(x_{n+1}, f(x_{n+1})), \ldots$ in the x–y plane. One gets each subsequent point by moving horizontally from the current point to the angle bisector $y = x$, and from there, vertically to the parabola $y = 4rx(1-x)$. Figures 3.6 and 3.7 show this construction, which is performed rather easily with *Mathematica*. The scaling laws (3.4) and (3.5) allow us to determine the universal Feigenbaum constant δ. To this end, we calculate the parameters R_l, at which there

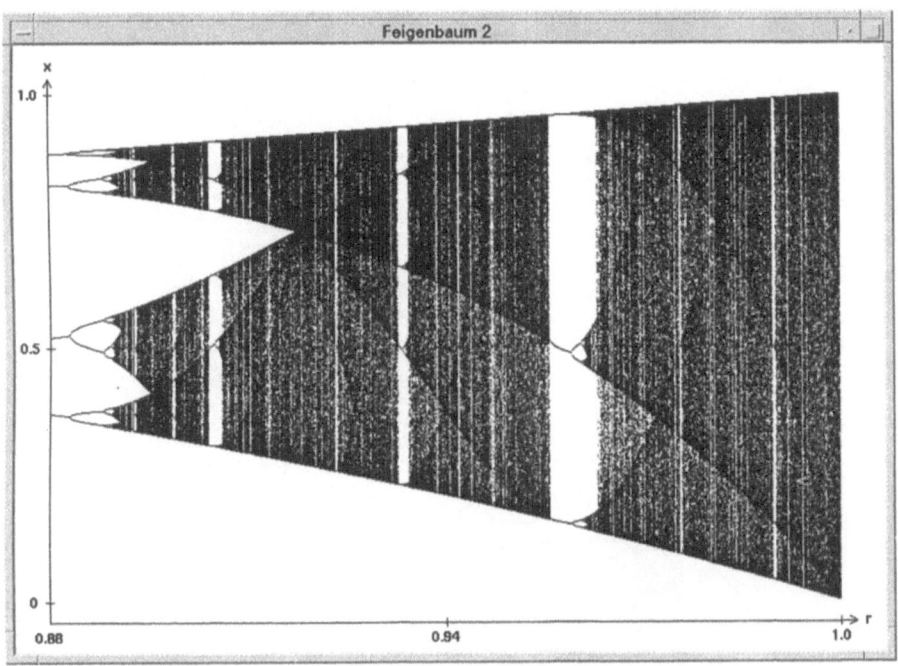

Fig. 3.4. A close-up from Fig. 3.3, for r between 0.88 and 1

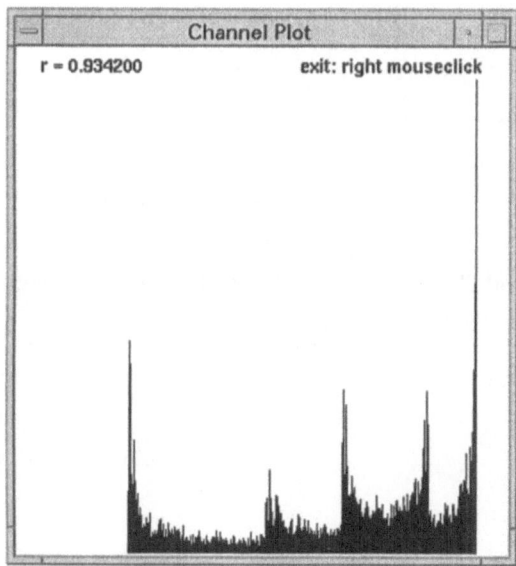

Fig. 3.5. Frequency distribution of the x_n values in the iteration of the parabola for a parameter value $r \simeq 0.934$, just below the window with period 5

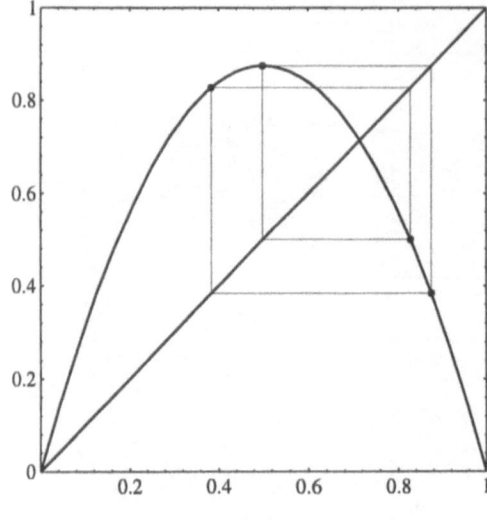

Fig. 3.6. Superstable four-cycle
with $r = 0.874640$

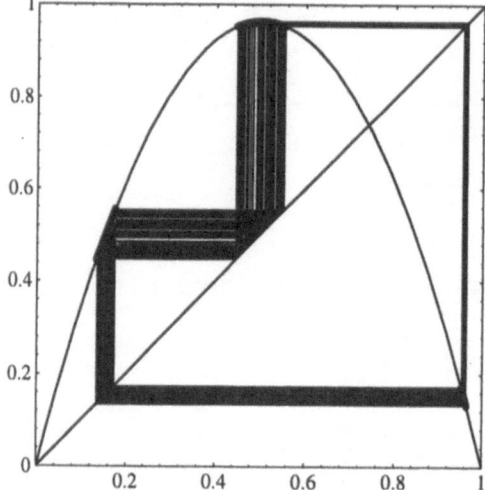

Fig. 3.7. Three chaotic bands for
$r = 0.9642$, just above the large
window with the three-cycle

are superstable orbits of period 2^l. Using the methods of symbolic dynamics
and inverse iteration we find the R_l with a precision of 30 digits, of which
only 12 are printed here:

$$R_{10} = 0.892486338871,$$
$$R_{11} = 0.892486401027,$$
$$R_{12} = 0.892486414339,$$
$$R_{13} = 0.892486417190,$$
$$R_{14} = 0.892486417801,$$
$$R_{15} = 0.892486417932.$$

It is worth noting that the last value is obtained by solving an equation with a 32 768-fold nested root!

Using

$$\delta_l = \frac{R_{l-1} - R_{l-2}}{R_l - R_{l-1}},$$

we find the following approximate values for δ:

$\delta_{10} = 4.669201134601$,

$\delta_{11} = 4.669201509514$,

$\delta_{12} = 4.669201587522$,

$\delta_{13} = 4.669201604512$,

$\delta_{14} = 4.669201608116$,

$\delta_{15} = 4.669201608892$.

From this we estimate that we have calculated δ to 9 digits. The extrapolation of δ_l for $l \to \infty$ is left as an exercise to the reader.

Finally, we want to investigate a seemingly simple question: given $x_0 = 1/3$ and $r = 97/100$, what is the value of x_{100}? This seems to be easy to answer; with

```
h[x_]:= 97/25 x(1-x)
```

the command

```
Nest[h, N[1/3],100]
```

yields the value $x_{100} = 0.144807$. This result is wrong, however; the correct value is $x_{100} = 0.410091$.

The solution to this discrepancy lies in the chaotic behavior of the logistic map for the given parameter value. The Lyapunov exponent, defined in (3.6) to (3.8), is positive. Therefore, small inaccuracies in the calculation lead to large errors after just a few iterations. Owing to the command N[1/3], *Mathematica* calculates everything to machine precision, which is 16 digits in our example. In each step of the iteration, the result is rounded; therefore the deviation from the true value, which slightly increases in each step, cannot be tracked. By using the format N[1/3,prec], on the other hand, one can specify the precision via prec. Now *Mathematica* reduces the calculated precision after each step until, finally, the variable x_n is set to the value 0 when only one digit is left. The command Precision[x] asks for the precision of x. Figure 3.8 shows the result. For prec=16, each result is rounded, so the precision does not seem to change. The improved calculation shows, however, that the correct result is obtained only if each step is calculated with a precision of more than 60 digits. We do not know the reason for the linear relation between the precision of the calculation and that of the result.

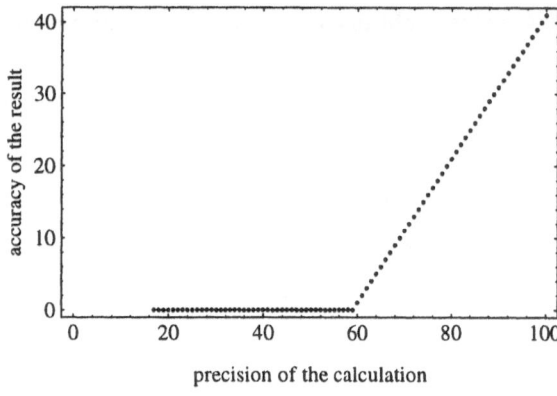

Fig. 3.8. The accuracy of x_{100} for $r = 97/100$ as a function of the precision with which the calculation is performed. The vertical axis gives the number of digits to which *Mathematica* calculates the result of the iteration

Exercises

1. Calculate the Lyapunov exponent for the logistic map (3.1) as a function of the parameter r.
2. In each r interval in the chaotic region, there are windows with superstable periodic orbits. Any such orbit starts with the symbolic dynamic CRL. Therefore, there are at most four possibilities for a five-cycle: $CRLRR$, $CRLRL$, $CRLLR$ and $CRLLL$. One of these possibilities is not realized, however. Calculate the parameter values of the remaining three orbits of period five.

Literature

Crandall R.E. (1991) Mathematica for the Sciences. Addison-Wesley, Redwood City, CA

Bai-Lin H. (1984) Chaos. World Scientific, Singapore

Jodl H.-J., Korsch H.J. (1994) Chaos: A Program Collection for the PC. Springer, Berlin, Heidelberg, New York

Ott E. (1993) Chaos in Dynamical Systems. Cambridge University Press, Cambridge, New York

Peitgen H.O., Saupe D. (1988) The Science of Fractal Images. Springer, Berlin, Heidelberg, New York

Schroeder M. (1991) Fractals, Chaos, Power Laws: Minutes from an Infinite Paradise. W.H. Freeman, New York

Schuster H.G. (1995) Deterministic Chaos: An Introduction. VCH, Weinheim, New York

3.2 Chain on a Corrugated Surface: The Frenkel–Kontorova Model

If two physical forces compete with one another, the resulting state can have intriguing properties. We have already seen in the case of the Hofstadter butterfly how two competing length scales – the scale generated by the lattice and the one generated by the magnetic field – resulted in a surprisingly complex energy spectrum. Now we want to look at a simple mechanical problem with classical equilibrium conditions: a chain of particles which interact with one another through linear forces and are subject to an additional periodic potential. We investigate only the rest positions of the chain elements. Both the spring forces and the external forces want to impose a distance of their own on the particles. It turns out that the stable states of the chain are described by iterations of a nonlinear, two-dimensional function, whose orbits can take a large variety of forms.

Physics

Our system is a one-dimensional arrangement of an infinite number of point-like masses whose rest state shall be characterized by the positions $x_n (n \in \mathbb{Z})$. The forces acting on the particles shall result from an external periodic potential $V(x)$, on the one hand, and from an interaction potential $W(x_n - x_{n-1})$ between neighboring particles, on the other. Then the energy of a long segment of the chain containing the sites $x_{M+1}, x_{M+2}, \ldots, x_N$ is given by

$$H_{MN} = \sum_{n=M+1}^{N} [V(x_n) + W(x_n - x_{n-1})] . \tag{3.10}$$

All lengths x are measured in units of the period of the external potential, i.e.,

$$V(x+1) = V(x) . \tag{3.11}$$

In the rest position, all forces on any given particle cancel each other. Therefore, we have

$$0 = \frac{\partial H_{MN}}{\partial x_n} = V'(x_n) + W'(x_n - x_{n-1}) - W'(x_{n+1} - x_n) . \tag{3.12}$$

With the definition

$$p_n = W'(x_n - x_{n-1}) \tag{3.13}$$

this leads to

$$\begin{aligned} p_{n+1} &= p_n + V'(x_n) , \\ x_{n+1} &= x_n + (W')^{-1}(p_{n+1}) , \end{aligned} \tag{3.14}$$

or, transformed to the inverse recursion,

$$x_{n-1} = x_n - (W')^{-1}(p_n) ,$$
$$p_{n-1} = p_n - V'(x_{n-1}) . \tag{3.15}$$

From (x_0, p_0) – equivalent to this is the specification of (x_{-1}, x_0) – one obtains the pairs (x_1, p_1) and (x_{-1}, p_{-1}), and so on. Therefore, the system of equations (3.14) describes a nonlinear iteration in the x–p plane. Each initial point (x_{-1}, x_0) uniquely determines a state $\{\dots, x_{-2}, x_{-1}, x_0, x_1, x_2, \dots\}$ of the entire chain, for which each particle is in its rest position.

The functional determinant of the mapping $(x_n, p_n) \to (x_{n+1}, p_{n+1})$ has the value one, $|\partial(x_{n+1}, p_{n+1})/\partial(x_n, p_n)| = 1$; therefore, the function is area-preserving. Even though a section of the x–p plane is deformed by the function, its image has the same area as the original section. Area-preserving flux in phase space is known from classical mechanics (Liouville's theorem), and, indeed, one can regard (3.14) as snapshots (Poincaré sections, see Sect. 4.2) of a continuous dynamic in three-dimensional phase space.

So far, the potentials V and W have not been specified, but now they shall be restricted to an external cosine potential and linear spring forces:

$$V(x) = \frac{K}{(2\pi)^2}[1 - \cos(2\pi x)] ,$$
$$W(\Delta) = \frac{1}{2}(\Delta - \sigma)^2 . \tag{3.16}$$

K determines the strength of the external potential, and σ is the distance between the particles for $K = 0$ in units of the period of the potential. We slightly modify (3.13), which yields $p_n = x_n - x_{n-1} - \sigma$, by setting

$$p_n = x_n - x_{n-1} . \tag{3.17}$$

Then, instead of (3.14), we obtain

$$p_{n+1} = p_n + \frac{K}{2\pi}\sin(2\pi x_n) ,$$
$$x_{n+1} = x_n + p_{n+1} . \tag{3.18}$$

This mapping, which is termed "standard map" or "Chirikov map" in the literature, has been investigated in many publications. The parameter σ no longer shows up in the iteration, only in the energy,

$$H_{MN} = K \sum_{n=M+1}^{N} \frac{1}{(2\pi)^2}(1 - \cos 2\pi x_n) + \sum_{n=M+1}^{N} \frac{1}{2}(p_n - \sigma)^2 . \tag{3.19}$$

For given parameters K and σ, the average energy per particle is

$$h = \lim_{N-M \to \infty} \frac{1}{N-M} H_{MN} , \tag{3.20}$$

which is a function of x_0 and p_0. The ground state is determined by the point (x_0, p_0) with the smallest energy.

An interesting property of a state is described by the so-called winding number w, which is defined as the mean distance between neighboring particles:

$$w = \langle x_{n+1} - x_n \rangle = \lim_{N-M \to \infty} \frac{x_N - x_M}{N - M} . \tag{3.21}$$

If a state is periodic relative to the lattice (a commensurate state), there are integers P and Q, with P and Q relatively prime, and $x_{n+Q} = x_n + P$. This leads to

$$w = \frac{x_{n+Q} - x_n}{Q} = \frac{P}{Q} . \tag{3.22}$$

Therefore, a commensurate state has a rational mean particle distance w. A state with an irrational w, on the other hand, does not lock in to the lattice, which affects some other physical properties as well. Here, as in the case of the Hofstadter butterfly, physics distinguishes between rational and irrational numbers!

Algorithm

Programming the mapping (3.18) is very easy. Since we are dealing with numerical manipulations, we define pi=N[Pi] in *Mathematica* and define the function t by

```
t[{x_,p_}] = {x + p + k/(2 pi) Sin[2 pi x],
              p + k/(2 pi) Sin[2 pi x]}
```

Iterations are easily executed via Nest or NestList:

```
list[x0_,p0_] := NestList[t,{x0,p0},nmax]
```

The result of list[ ] is a list of nmax+1 points in the x–p plane, which describes the physical state for a given set of initial values x_0 and $p_0 = x_0 - x_{-1}$. Thus, after specifying the positions x_0 and x_{-1} of two neighboring particles, list[] yields the positions of the other particles.

We obtain an intuitive picture of the resulting states by placing the particles on the curve $V(x)$ to get the coordinates $\{x_n, V(x_n)\}$, and plotting these points with ListPlot. The coordinates x_n are extracted from the result of list[...] with the command

```
xlist[x0_,p0_] := Map[First,list[x0,p0]]
```

For the pendulum (Sect. 1.2) we have already seen the usefulness of investigating the phase-space diagram in the x–p plane rather than the function x_n (in the case of the pendulum it was the $\dot\varphi$–φ diagram instead of $\varphi(t)$). To this end, we only have to display each of the variables x and p in the interval $[0, 1]$, since the function (3.18) yields the same values (modulo 1) if the number 1 is added to x_n or p_n.

The modulo function can be applied to a list and made to act on the individual coordinates via Map:

```
tilde[{x_,p_}]:= {Mod[x,1],Mod[p,1]}
listt[x0_,p0_]:= Map[tilde,list[x0,p0]]
```

The phase-space diagram is generated with ListPlot. The energy and winding number are easily calculated from the state 111 = list[x0,p0]:

```
de[{x_,p_}]:=k/(2 pi)^2 (1-Cos[2 pi x])+(p-sigma)^2/2
```

```
h[x0_:.0838,p0_]:=
        ( 111=list[x0,p0];
          112=Map[de,111];
          Apply[Plus,112]/Length[112])
```

```
wind[x0_:.0838,p0_]:=
        ( w1=xlist[x0,p0]; (w1[[-1]] - w1[[1]])/nmax)
```

Again, the calculation in *Mathematica* takes a very long time. Therefore, we want to regenerate the phase-space diagram directly on the monitor using a C program. Additionally, the energy and winding number are to be calculated. The starting point is randomly selected by the program.

In C, the mapping (3.18) is written as

```
pnew = p+K/2./pi*sin(2.*pi*x);
xnew = x+pnew;
```

Properly, the expressions K/2./pi and 2.*pi should not be reevaluated each time in this inner loop, where these commands are executed very frequently, but should be replaced by constants. In our program, however, the graphics require so much cpu time that we do not make this effort to optimize the algorithm. Now the x–p values are transformed into image coordinates:

```
xs = fmod(xnew,1.)*xmax+10;
ys = fmod(pnew,1.)*ymax+100;
```

and a point is plotted at the position (xs,ys), e.g., on the PC via

```
putpixel(xs,ys,color);
```

For each call, a new color is selected:

```
color = random(getmaxcolor())+1;
```

In this way, a display with different orbits in different colors is generated on the screen, each one representing one state.

Results

Figure 3.9 shows the results of iterating (3.18) for $K = 1$. The numbers (x_0, p_0) were selected randomly in the unit square, and for each starting point 10 000 (x_n, p_n) pairs were calculated and plotted with the C program. This picture is independent of the ratio σ of competing length scales, since

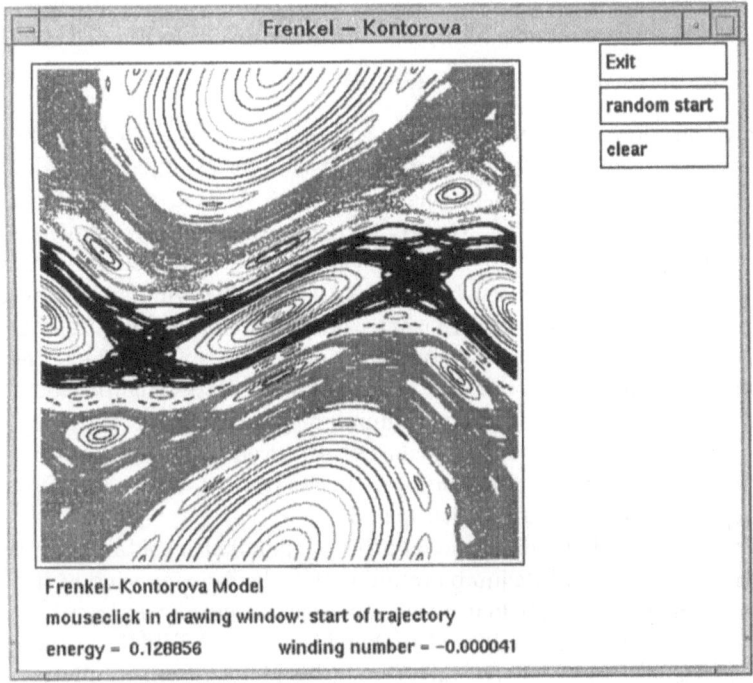

Within the figure window:

Frenkel — Kontorova

Exit

random start

clear

Frenkel–Kontorova Model

mouseclick in drawing window: start of trajectory

energy = 0.128856   winding number = −0.000041

Fig. 3.9. Orbits in the x–p plane. On the workstation, a new starting point can be selected by a mouse click and the resulting orbit is plotted in a new color

this parameter is not used in the iterated function. Each orbit has a winding number w, which depends on the starting point and the periodic potential, but not on σ. The equilibrium constant σ shows up only in the energy per particle:

$$h = \frac{1}{N} \sum_{n=1}^{N} \left[\frac{K}{(2\pi)^2} \left(1 - \cos\left(2\pi x_n\right)\right) + \frac{1}{2} \left(p_n - \sigma\right)^2 \right] .$$

There are three different kinds of orbits in Fig. 3.9: zero-, one-, and two-dimensional ones. The zero-dimensional orbits consist of a finite number of points, which are found, for example, in the centers of the islands. They are the commensurate states with rational winding numbers $w = P/Q$ and period Q. P indicates the order in which the orbit loops through the Q points (x_n, p_n). Around these commensurate states, one sees closed orbits, i.e., one-dimensional ones. These states look almost the same as their commensurate relatives in the centers of the orbits, but each x_n oscillates about the corresponding commensurate value with a frequency that is incommensurate with the period of the potential.

We want to investigate this for $w = 2/5$ with *Mathematica*. First, by using FindRoot, we solve the equation

```
Mod[t[t[t[t[t[{x0,p0}]]]]],1] == {x0,p0}
```

The initial value, however, needs to be specified rather close to the solution, so the routine can find that solution. We obtain, for example, the value $(x_0, p_0) = (0.0838255, 0.336171)$ as one of the $Q = 5$ fixed points of the five-fold iterated function t[x,p]. Figure 3.10 then shows an incommensurate state with $(x_0, p_0) = (0.08, 0.34)$, i.e., close to the commensurate state occupying the centers of the five islands. It has the same winding number $w = 0.4$ and a slightly higher energy than the periodic state. Figure 3.11 shows the quasiperiodic state in configuration space. For the particle at the potential minimum, it can be seen that its position oscillates slightly about the minimum. Approximately the same minimum position is reached again after $Q = 5$ iterations. The oscillation can be seen even better in Fig. 3.12, where the x_n are plotted as a function of n. In addition to the period with $Q = 5$, there is another oscillation with $Q \simeq 30$.

Next to the zero- and one-dimensional orbits, Fig. 3.9 also contains states that fill entire areas. Such orbits jump around within the drawing window in a chaotic manner. Some orbits obviously span the entire range of x and p values. The latter is not true for small values of K. For $K < K_c = 0.971635 \ldots$, there are one-dimensional orbits that traverse the entire picture horizontally (so-called KAM orbits). It can be shown that chaotic orbits cannot cross the one-dimensional ones. Consequently, they are limited to a p interval between two KAM orbits. At K_c, the last KAM orbit vanishes – namely, the one with the "most irrational" winding number, the inverse golden ratio $w = 2/(1 + \sqrt{5}) = 0.618 \ldots$.

We are, however, not only interested in the properties of the standard map, but are also looking for the state with the lowest energy for a given value of σ. In principle, we can find the minimum by calculating $h(x_0, p_0)$ for different values of (x_0, p_0). In practice, though, there are better methods,

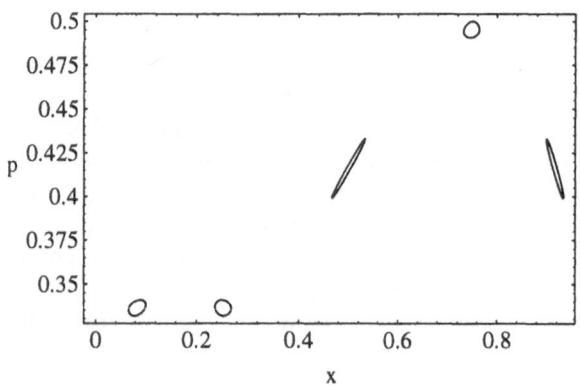

Fig. 3.10. Quasiperiodic orbit in phase space (x, p) (modulo 1) resulting from the initial values $(x_0, p_0) = (0.08, 0.34)$

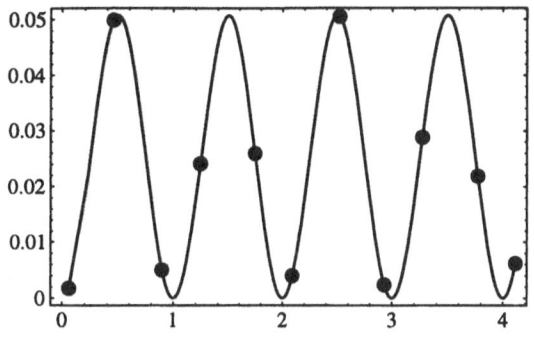

Fig. 3.11. The same quasi-periodic state as in Fig. 3.10, overlaid on top of the potential $V(x)$

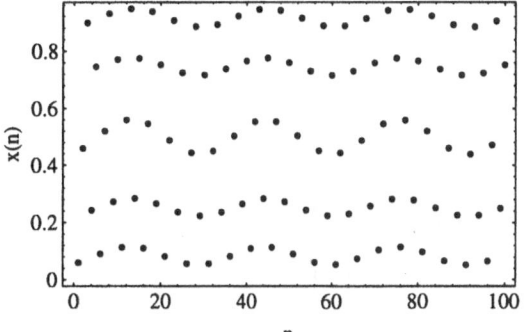

Fig. 3.12. The displacement x (modulo 1) of the nth particle is plotted for the same function as in Figs. 3.10 and 3.11

which we will not discuss here (see, e.g., Griffiths). Figure 3.13 shows $h(x_0, p_0)$ in a slice of the x_0–p_0 plane for x_0 fixed at the value 0.0838 and $\sigma = 2/5$. The corresponding winding numbers w can be seen in Fig. 3.14. In the interval shown, h reaches a minimum at $p_0 \simeq 0.35$. If we want to find the global minimum of the energy, we have to scan the function $h(x_0, p_0)$ in the entire x_0–p_0 plane. At the minimum, w appears to be "locked" into the value $w = \sigma = 0.4$. If the precise minimum of $h(x_0, p_0)$ and the corresponding winding number w are determined for all values of σ, $w(\sigma)$ exhibits a fascinating behavior, which is also called the *devil's staircase*: w "locks" into each rational number P/Q in steps, whose width decreases as Q increases. Consequently, $w(\sigma)$ has an infinite number of steps, and between any two steps there is again an infinite number of steps. Nevertheless, $w(\sigma)$ is continuous; for all irrational winding numbers there is a σ value and a corresponding ground state. This behavior can be proven mathematically.

The function $w(\sigma)$, which resists all imagination, results from the competition of two lengths. The commensurate ground states belonging to the rational values of w correspond to individual points in the x–p plane, while it can be shown that the incommensurate ground states do not correspond to area-covering chaotic orbits; rather they correspond exclusively to lines. As a side note, incommensurate states can be dragged across the cosine potential without changing their energy.

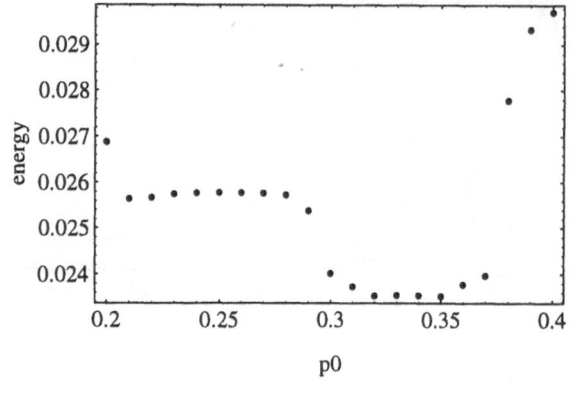

Fig. 3.13. Energy of a force-free state with $\sigma = 0.4$ and initial value $x_0 = 0.0838$ as a function of p_0

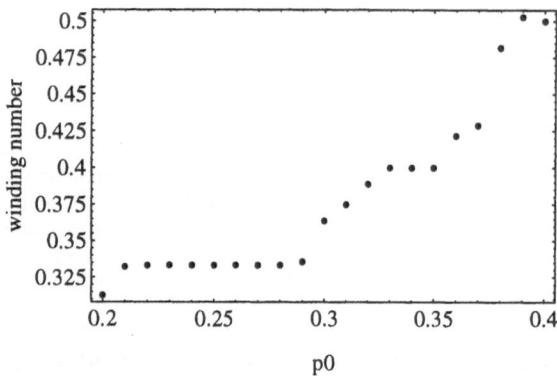

Fig. 3.14. Mean distance w between neighboring particles for the force-free states shown in Fig. 3.13

Exercise

For different values of K from (3.18), plot orbits in the x–p plane. For $K = 0$, you should be able to draw them by hand. Observe how the closed orbits, the so-called KAM orbits, turn into islands or chaotic orbits as K increases; calculate the winding numbers of the remaining KAM orbits. Attempt to find the last closed orbit just before K_c.

Literature

Jodl H.-J., Korsch H.J. (1994) Chaos: A Program Collection for the PC. Springer, Berlin, Heidelberg, New York

Ott E. (1993) Chaos in Dynamical Systems. Cambridge University Press, Cambridge, New York

Schuster H.G. (1995) Deterministic Chaos: An Introduction. VCH, Weinheim, New York

Griffiths R.B. (1990) Frenkel–Kontorova Models of Commensurate–Incommensurate Phase Transitions. In: van Beijeren H. (Ed.) Fundamental

Problems in Statistical Mechanics VII. Elsevier, Amsterdam, New York, 69–110

Greene J.M. (1979) A Method for Determining a Stochastic Transition. J. Math. Phys. 20:1183

3.3 Fractal Lattices

Consider a simple game of dice and a simple geometric iteration. Both lead to a strange structure that is less than an area, but more than a line – a so-called fractal with a dimension $D = \log 3/\log 2 = 1.58\dots$. Surprisingly, fractals are formed very frequently in nature. Coastlines, mountain ranges, blood vessels, courses of rivers, variations of stock prices, and water levels in rivers, can all be described by a fractional dimension. In addition to the geometric game, in this section we will learn about aggregates and percolation clusters as examples of these remarkable self-similar objects.

We first want to deal with the definition of the fractal dimension. Then, we will present two programs that use different methods to construct the Sierpinski gasket, which is often presented as a standard example of fractals.

Physics

The spatial dimension D of an object can be determined from the relation between the number of its components that make up its mass M and its linear dimension L:

$$M \propto L^D . \tag{3.23}$$

Thus, by comparing the masses of two similar cubes with sides L and $2L$, one obtains

$$\frac{M_2}{M_1} = \left(\frac{2L}{L}\right)^D = 2^D \tag{3.24}$$

with $D = 3$. For squares, the result is obviously $D = 2$, and for lines $D = 1$. L can also be the radius of spheres or disks, or any other characteristic length of the object under consideration; of course the mass is proportional to the number of particles. In any case we have

$$D = \lim_{L \to \infty} \frac{\log M}{\log L} . \tag{3.25}$$

There is yet another way of determining the dimension D. For this method, we completely cover the object with small cubes whose edges have the length ε. Let $N(\varepsilon)$ be the smallest number of these cubes needed. Then $N(\varepsilon) \propto \varepsilon^{-D}$, and consequently

$$D = -\lim_{\varepsilon \to 0} \frac{\log N(\varepsilon)}{\log \varepsilon} . \tag{3.26}$$

To cover a cube, we need $(L/\varepsilon)^3$ boxes, whereas the number is $(L/\varepsilon)^2$ for a square and $(L/\varepsilon)^1$ for a line.

So, if we want to determine the length of a border on a map, for example, we can take a small ruler of length ε and determine the number $N(\varepsilon)$ of steps we need to measure the borderline. Then, we would expect

$$L = \varepsilon N (\varepsilon) \tag{3.27}$$

to be a measure of that length, as we think of a border as a one-dimensional object with $D = 1$. By this method, the Spaniards have obtained a value of 987 km for the length of their country's border with Portugal, whereas the Portuguese have determined the same border to be 1214 km long. A smaller ruler ε apparently requires more steps than L/ε to cover the border, therefore, D cannot be equal to 1. Consequently, (3.27) will yield a value for L that depends on ε. A more detailed analysis of the data reveals the following: A border is a fractal with $1 < D < 2$; therefore, (3.27) yields $L \propto \varepsilon^{1-D}$, i.e., a length that seems to diverge as $\varepsilon \to 0$. The border does not have a well-defined length at all! Of course, this can only be true for scales ε that are larger than the smallest (e.g., the surveyor's rods used) and smaller than the largest length involved (e.g., the size of Portugal).

Another simple example, in which the fractal dimension D does not agree with the embedding dimension d, that is, the dimension of the space in which the object is located, is a random motion in d-dimensional space. This *random walk*, as it is commonly called, is also discussed as a simple model of a polymer molecule that consists of a large number of monomers. We assume in this discussion that the distance between adjacent monomers is constant, but that the angles between three consecutive monomers are independent and randomly distributed. Then, the vectors r_i connecting adjacent monomers are random vectors, whose properties are easily calculated. If $\langle \ldots \rangle$ denotes the average over many molecular configurations, we have:

$$\langle r_i \rangle = 0 , \quad \langle r_i \cdot r_j \rangle = \langle r_i \rangle \cdot \langle r_j \rangle = 0 \text{ for } i \neq j , \quad \text{and} \quad \langle r_i^2 \rangle = a^2 ,$$

where a is the distance between adjacent monomers. For the vector R between the beginning and end of a molecule consisting of $N + 1$ monomers, we have

$$R = \sum_{i=1}^{N} r_i$$

and therefore

$$\langle R \rangle = \sum_{i=1}^{N} \langle r_i \rangle = 0$$

and

$$\langle R^2 \rangle = \sum_{i=1}^{N} \sum_{j=1}^{N} \langle r_i \cdot r_j \rangle = \sum_{i=1}^{N} \langle r_i^2 \rangle + \sum_{i \neq j} \langle r_i \cdot r_j \rangle = Na^2 .$$

Since the mass M increases proportionally to the number of N of monomers and the linear dimension L is proportional to $\sqrt{\langle R^2 \rangle}$, this means that

$$M \propto L^2$$

and consequently, using (3.23), $D = 2$. Such a random walk, therefore, is always a two-dimensional object, independent of the dimension d of the space in which it is generated. The fractal dimension is reduced, however, if the size of the individual monomers is taken into account. We will discuss this effect in Sect. 5.4.

Algorithm and Result

Now we want to take a closer look at the two games mentioned at the beginning of this section. First, the algorithm for the game of dice:

- Select three points p_1, p_2, and p_3 and a starting point q_0 arbitrarily distributed in the plane.
- Starting from the point q_n, construct the next point q_{n+1} by the following method: randomly (e.g., by casting a die) pick one of the three points p_i. Then determine the center of the line connecting q_n and p_i:

$$q_{n+1} = \frac{(q_n + p_i)}{2} .$$

- Iterate the equation above ad infinitum.

What pattern do the points $\{q_0, q_1, q_2, \ldots\}$ generate in the plane? The algorithm, which we write in C for the sake of increased processing speed, is easy to program. We define each point as a structure with two integer variables x and y, representing the pixel coordinates on the screen.

```
struct{int x, int y} qn={20,20}, pw,
        p[3]={{10,10},{MAXX-10,10},{MAXX/3,MAXY-10}} ;
```

The initial values are specified right in the type declaration. Now in each step of the iteration one of the three points p[i] is selected by generating a random number $i \in \{0, 1, 2\}$,

```
pw = p[(3*rand())/RAND_MAX]
```

and a new point is calculated:

```
qn.x = (pw.x + qn.x)/2 ;
qn.y = (pw.y + qn.y)/2 ;
```

Then a pixel is plotted at that position, e.g., on the PC via

```
putpixel(qn.x, qn.y, WHITE);
```

Figure 3.15 shows the result. Surprisingly, the random motion of the point q leads to an obviously regular structure consisting of triangles nested inside each other. The structure is self-similar: all triangles look the same, independent of their size (of course, this is only true for triangles that are much larger than the size of the pixels). But the structure is nearly empty, as each triangle has an infinite number of holes at any scale. After just a few iterations, any point we start with ends up in this strange grid; thus, the framework, also called the *Sierpinski gasket*, is an attractor for the dynamics of the entire plane.

The second geometric game goes as follows:

- Start with a triangle,
- remove a triangle from its center in such a way that three identical triangles remain,
- iterate this ad infinitum for all triangles remaining after each step.

To take advantage of the simple graphics commands, we want to program this in *Mathematica*. Each triangle is described by a list of three points in the plane. The initial triangle is given by:

```
list={{{0.,0.},{.5,N[Sqrt[3/4]]},{1.,0.}}}
```

Each triangle is tripled by first scaling it down by a factor $1/2$ and shifting it along two of its edges by the length of the respective edge:

```
triple[d_]:= Block[ {d1,d2,d3},
                d1={d[[1]],(d[[2]]+d[[1]])*.5,
                         (d[[3]]+d[[1]])*.5};
                d2=d1+Table[(d1[[3]]-d1[[1]]),{3}];
```

Fig. 3.15. Points in the plane, generated by the game of dice described in the text. The corners of the Sierpinski triangle are given by the points p[0], p[1], and p[2]

```
d3=d1+Table[(d1[[2]]-d1[[1]]),{3}];
{d1,d2,d3}
]
```

The command Map is used to repeatedly apply this function to the entire list
of triangles. Finally, the command Polygon turns the triangles into graphics
objects which are drawn by Show:

```
plot1:= Block[ {listtwo,plotlist},
                listtwo=Map[triple,list];
                list=Flatten[listtwo,1];
                plotlist=Map[Polygon,list];
                Show[Graphics[plotlist],
                    AspectRatio -> Automatic]
         ]
```

Figure 3.16 shows the result. We obviously obtain the same structure as in
the previous game of dice, but now, we can easily convince ourselves that the
self-similar framework is a fractal in the limit of infinitely many iterations.
If we cut in half an edge of the triangle of length L, its mass is obviously
reduced by a factor 3. After t steps, we have

$$M_t = 3^{-t}M , \quad L_t = 2^{-t}L .$$

This leads to

$$D = \lim_{t \to \infty} \frac{\ln M_t}{\ln L_t} = \lim_{t \to \infty} \frac{(-t)\ln 3 + \ln M}{(-t)\ln 2 + \ln L} = \frac{\ln 3}{\ln 2} .$$

Fig. 3.16. By repeated removal of triangles, a fractal of dimension $D \simeq 1.58$ is
generated

Thus, the Sierpinski gasket has the fractional dimension $D = 1.58\ldots$. While it does not fill an area, it is indeed more than a line. Its density $\rho = M/L^2$ is zero, and any triangle, however small it may be, looks the same as a big one.

It is remarkable that nature uses fractals in order to effectively supply blood to every part of the body, for example, or to generate spacesaving direct connections between 10^{11} communicating nerve cells in the brain. So far, technology has not been able to make much use of fractals. The book by M. Schroeder, however, does describe the use of fractal reflectors in acoustics.

Exercise

A Koch curve is defined by the following iteration: The central third of a line is cut out and replaced by a protrusion whose two edges each have the same length as the segment just removed, as shown in the following drawing.

Start with an equilateral triangle, and iterate the construction above as often as possible. As a result, you will obtain the Koch snowflake.

Literature

Ott E. (1993) Chaos in Dynamical Systems. Cambridge University Press, Cambridge, New York

Peitgen H.O., Saupe D. (1988) The Science of Fractal Images. Springer, Berlin, Heidelberg, New York

Schroeder M. (1991) Fractals, Chaos, Power Laws: Minutes from an Infinite Paradise. W.H. Freeman, New York

Schuster H.G. (1995) Deterministic Chaos: An Introduction. VCH, Weinheim, New York

Wagon S. (1991) Mathematica in Action. W.H. Freeman, New York

3.4 Neural Networks

Can modern computers simulate the human brain? Anybody who thinks about it will probably reach the conclusion that the answer to this bold question must be a definite no. While a lot of facts are known nowadays about the way our brain works, it is still unknown how the exchange of information between 10^{11} cells (neurons) and their 10^{14} contact points (synapses) can result in processes as commonplace to us as learning, recognition, and thinking.

We know little, yet at present, a few thousand scientists and engineers are trying to develop computer programs that learn from the brain's architecture, as well as from its function. Such algorithms and their realizations in hardware are called *neural networks* and *neural computers*, respectively. Indeed, they have features that are clearly different from those of present-day computers.

A neural network is not fed a computer program; instead, it learns from examples by adjusting the strengths of its synapses to them. After the learning phase, it can generalize, i.e., it has deduced rules from the examples. The information obtained is not stored in numbered drawers, as in a (classical) computer, but is distributed over the entire network of synapses. The logic is not a strict "yes or no", but a vague "more or less."

Many impressive applications of neural networks have been demonstrated in recent years: pronouncing English text, playing backgammon, recognizing digits, recognizing engine defects by their noises, judging the creditworthiness of bank customers, predicting stock prices, etc. It turns out that neural networks can often compete with other, more complicated methods. This is achieved with learning rules that are surprisingly simple and universal.

The simplest neural network, the so-called *perceptron*, has already been used and mathematically investigated in the sixties. We want to present and program it here. Our perceptron will learn to predict the keyboard entries 1 and 0. Even if the reader tries selecting the two keys in a completely random way the network will, on average, predict better than 50% correct. While this ability is still rather moderate, a computer with a five-line algorithm written in C can learn to predict human behavior; this possibility should give us pause after all.

Physics

The perceptron consists of an input layer of "neurons" S_i, an additional layer of N "synaptic" weights w_i $(i = 1, \ldots, N)$, and an output neuron S_0, which is directly connected to all input neurons via the synapses. As with real nerve cells, the element S_0 reacts to the sum of the activities of those neurons which can directly act upon S_0 via the synaptic connections. As early as 1943, this rather complex, time-dependent process was described by a simple mathematical formula:

$$S_0 = \text{sign} \left(\sum_{j=1}^{N} S_j w_j \right) . \tag{3.28}$$

Here, any neuron can have just two states: it is quiescent $(S_i = -1)$, or it sends pulses $(S_i = 1)$. The coefficients $w_i \in \mathbb{R}$ model the strength with which the incoming signal from the neuron S_i is converted, by the synapse, to an electric potential in the nucleus of S_0. The real number w_i describes complex biochemical processes; for example, a synapse can have an inhibiting

($w_i < 0$) or a stimulating effect ($w_i > 0$). If the sum of the potentials exceeds the threshold value 0, the neuron fires ($S_0 = +1$), otherwise it is quiescent ($S_0 = -1$).

This is the biological motivation for (3.28). Mathematically speaking, this equation defines a Boolean function $\{+1, -1\}^N \rightarrow \{+1, -1\}$, which classifies each of the possible inputs $S = (S_1, S_2, \ldots, S_N)$ by $+1$ or -1. There are 2^N different inputs. In general, any of these inputs can be labeled by $+1$ or -1, so there are $2^{(2^N)}$ Boolean functions; for $N = 10$, one obtains the unimaginably large number 10^{308}.

The perceptron characterized by (3.28) only defines specific Boolean functions, however. Geometrically speaking, S and $w = (w_1, w_2, \ldots, w_N)$ are vectors in N-dimensional space. Equation (3.28) separates the outputs $S_0 = 1$ and $S_0 = -1$ by the $(N-1)$-dimensional hyperplane $w \cdot S = 0$. Therefore, the perceptron is also called a linearly separable Boolean function. It can be shown that their number is less than $2^{(N^2)}$, which, for $N = 10$, still amounts to 10^{30}.

What is the meaning of learning and generalizing for the perceptron described by (3.28)? A neural network is supplied with neither rules nor programs, but examples. In our case, the examples consist of a set of input–output pairs (x_ν, y_ν). with $x_\nu = (x_{\nu 1}, x_{\nu 2}, \ldots, x_{\nu N})$, $y_\nu \in \{+1, -1\}$, and $\nu = 1, \ldots, M$, provided by an unknown function. As in the case of real nerve cells, our network learns by "synaptic plasticity", i.e., it adjusts the strength of its weights w_i to the examples. In the ideal case, the perceptron can correctly classify all inputs after the learning phase:

$$y_\nu = \text{sign}\,(w \cdot x_\nu) \ . \tag{3.29}$$

Effective learning rules for the perceptron have been known since the sixties. D. Hebb had already postulated in 1949 that synapses adjust themselves to the activities of their input and output neurons. Mathematically, this statement was later expressed in the form

$$\Delta w = \frac{1}{N} y_\nu x_\nu \ . \tag{3.30}$$

Upon presentation of the νth example, each synaptic strength w_i changes slightly, by an amount proportional to the product of the input and output activities. If only one example (x, y) is to be learned, it is easy to see that synapses which fulfill (3.30) also fulfill (3.29) perfectly, which is the condition to be learned ($x_j \in \{+1, -1\}$):

$$w \cdot x = \sum_{j=1}^{N} \left(\frac{y}{N} x_j \right) x_j = \frac{y}{N} \sum_{j=1}^{N} x_j x_j = y \ . \tag{3.31}$$

This method, however, no longer works if N neurons are supposed to learn $M = \mathcal{O}(N)$ examples; in that case, (3.30) needs to be slightly modified. Any example is to be learned according to (3.30) only if it is not yet classified correctly by the current vector w:

$$\Delta w = \frac{1}{N} y_\nu x_\nu \text{ if } (w \cdot x_\nu) y_\nu \leq 0, \quad \Delta w = 0 \text{ otherwise.} \tag{3.32}$$

This is the well-known perceptron learning rule, whose convergence was mathematically proven in 1960 by Rosenblatt: if the M examples (x_ν, y_ν) can be classified correctly by a perceptron at all, then the algorithm (3.32) will terminate. Consequently, in this case the learning rule (3.32) will always find a weight w, one of many possibilities, which fulfills (3.29) for all examples.

Since this textbook emphasizes algorithms, we want to present the proof at this point. Using the definition $z_\nu = y_\nu x_\nu$, we are looking for a w with $w \cdot z_\nu > 0$ for $\nu = 1, \ldots, M$. According to our assumption, such a w_* with $w_* \cdot z_\nu \geq c > 0$ for all ν exists. The constant $c > 0$ corresponds to the smallest distance of the points z_ν from the hyperplane defined by w_*.

Now we start the algorithm with $w(t = 0) = 0$. The variable t is a counter of those learning steps for which the weights are changed, i.e., those steps for which we have $w(t) \cdot z_\nu \leq 0$. For each of these, the learning rule yields

$$w(t + 1) = w(t) + \frac{1}{N} z_\nu . \tag{3.33}$$

Vector multiplication of (3.33) by itself yields

$$(w(t+1))^2 = (w(t))^2 + \frac{2}{N} w(t) \cdot z_\nu + \frac{1}{N^2} (z_\nu)^2$$
$$\leq (w(t))^2 + \frac{1}{N} . \tag{3.34}$$

Note that $(z_\nu)^2 = \sum_j z_{\nu j}^2 = N$. Iteration of (3.34) from 0 to t, using $w(0) = 0$, leads to :

$$(w(t))^2 \leq \frac{t}{N} . \tag{3.35}$$

An estimate of the scalar product $w_* \cdot w(t+1)$ yields

$$w_* \cdot w(t+1) = w_* \cdot w(t) + \frac{1}{N} w_* \cdot z_\nu \geq w_* \cdot w(t) + \frac{c}{N} , \tag{3.36}$$

which, iterated from 0 to t, leads to

$$w_* \cdot w(t) \geq \frac{ct}{N} . \tag{3.37}$$

Now we insert (3.35) and (3.37) into Schwarz's inequality

$$(w(t))^2 (w_*)^2 \geq (w(t) \cdot w_*)^2 \tag{3.38}$$

and obtain

$$\frac{t}{N} (w_*)^2 \geq \frac{c^2 t^2}{N^2} , \tag{3.39}$$

or

$$t \leq N \frac{(w_*)^2}{c^2} = t_* . \tag{3.40}$$

This means that the number t of learning steps is limited; the algorithm terminates after no more than t_* steps. It only ends, though, if all examples are classified correctly. Thus, the perceptron rule can actually learn all learnable problems perfectly.

If a large number of examples is to be learned, the distance c from the hyperplane gets small and consequently the number t_* of learning steps gets very large. This means that each example possibly needs to be learned very frequently before, eventually, all of them are classified correctly.

How many examples can a perceptron learn at all? Obviously, this depends on the unknown function that generates the examples. If these come from another perceptron, then, according to the convergence theorem above, any number M can be learned. If, on the other hand, the classification bits y_ν are selected randomly, we can make exact statements again: for $M < N$, all examples can be learned perfectly. In the limit $N \to \infty$, this is even true (with probability 1) for $M < 2N$.

After the learning phase, the perceptron can generalize. This means that it classifies an input S, which it has not learned before, in the same way as the unknown function from which it has learned. If, for an input S, the "teacher" function yields the result S_0, then a measure g of the perceptron's ability to generalize is defined as follows:

$$g = \langle \Theta\left(S_0 w \cdot S\right)\rangle_S , \qquad (3.41)$$

where the average is taken over many inputs S. The value $g = 1/2$ is the result of random guesses and the perceptron can not generalize in this case. For $g > 1/2$, on the other hand, the perceptron has recognized a certain regularity in the examples learned, and has a probability g of agreeing with the "teacher."

We now want to use the perceptron for the purpose of time series analysis and make predictions for the next step. We assume we have a sequence of bits, e.g.,

$$F = (1, -1, -1, 1, 1, -1, -1, -1, 1, 1, 1, -1, 1, 1, -1, \ldots) . \qquad (3.42)$$

In each step, the perceptron probes a window of N bits,

$$x_\nu = (F_\nu, F_{\nu+1}, \ldots, F_{\nu+N-1}) , \qquad (3.43)$$

and makes a prediction for the bit $F_{\nu+N}$:

$$\tilde{F}_{\nu+N} = \text{sign}\left(w \cdot x_\nu\right) . \qquad (3.44)$$

Then the perceptron learns the example $(x_\nu, F_{\nu+N})$ according to the Rosenblatt rule (3.32), shifts the window to the right by one bit, and makes a new prediction for the input $F_{\nu+N+1}$. This is repeated and a hit frequency g is determined.

If the sequence F has a period M, then there are obviously M examples, and for $M < N$, the perceptron can learn this "rhythm" perfectly. A random sequence, on the other hand, will always give a hit frequency of $g = 50\%$.

We want to enter a bit sequence ourselves and test whether (and how well) a simple perceptron can recognize obvious or subconscious regularities in it.

Algorithm

To this end, we program the perceptron in C. The neuron activities are stored in the array `neuron[N]`, and the synaptic weights in `weight[N]`. The variable `input` receives the next bit. In each step, only the latest example is learned. The learning and prediction algorithm from (3.32) and (3.44) is simply:

```
while (1) {
```

1. Read input:

```
if (getch() == '1') input = 1; else input = -1; runs++;
```

2. Calculate the potential:

```
for (h=0., i=0; i<N; i++)  h += weight[i] * neuron[i];
```

3. Count the number of correct predictions:

```
if (h*input > 0) correct++;
```

4. Learn:

```
if(h*input < 0)
for (i=0; i<N; i++)
weight[i] += input*neuron[i]/(float)N;
```

5. Shift the window:

```
for (i=N-1; i>0; i--) neuron[i] = neuron[i-1];
neuron[0] = input;
} /* end of while */
```

As we can see, the essential part of the program `nn.c` really consists of no more than five lines. Everything else is initialization and graphic display on the screen. The hit frequency in the form `correct/runs*100.0` is displayed after each step, as are the values of the weights and the result of the prediction. Hitting `n` on the keyboard causes the count to be reset; `e` is used to end the program.

The version `nnf name` takes the inputs 1 and 0 from the file `name`. At this point, we want to show briefly how to pass arguments from the command line and how to read a file. The function `main` now has arguments:

```
main(int argc, char *argv[])
```

`argc` contains the number of words entered on the command line, in this case `argc=2`. The two array elements `argv[0]` and `argv[1]` contain the addresses of the program name (`nnf`) and the argument (`name`). If no name is entered, we want the program to prompt the user for it and read the answer. The corresponding file is then opened for read access.

```
FILE *fp;
char str[100];
if(argc==1)   { printf("Input file name?");
                 scanf("%s", str);
              }
   else        strcpy(str, argv[1]);
if((fp=fopen(str,"r"))==NULL)
                 { printf("File not found");
                   exit(1); }
```

The characters 1 and 0 are then read by the following code:

```
while(feof(fp)==NULL)
    {   switch(fgetc(fp))
        {   case'1': input=1; break;
            case'0': input=-1; break;
            default: continue;
        }
    }
    ...
```

A non-null value of **feof** indicates the end of the file, and **fgetc** reads the next character from the file, which has been assigned the pointer **fp** by **fopen**. The instruction **break** exits from the **switch** command, and **continue** jumps past the rest of the **while** loop, which contains steps two to five above. The program **nnf** only accepts 1 and 0 as valid inputs for a bit sequence, whereas **nn** converts the inputs 1 or "different from 1" to +1 and -1.

Result

The command **nn** graphically displays the perceptron on the screen (Fig. 3.17). The bit sequence from the keyboard input 1 or 0, represented by red and green dots respectively, can be seen for the last 20 steps, but the perceptron only probes the last 10 bits and makes a prediction for the next input. Of course, that prediction is not shown; the result of the comparison of the two is not printed until after the input. The learning process can be observed in the change of the synaptic weights, which are drawn in the form of violet bars. The hit frequency g is the principal result; $g > 50\%$ means that the perceptron has recognized a structure in the bit sequence.

First, we type in a rhythm, e.g.,

1010101010 ... .

After a few steps the network has learned this pattern and makes only correct predictions. If the sequence is switched to a different period, e.g.,

111001110011100 ... ,

the perceptron learns this sequence perfectly in just a few steps as well. Of course, a return to the previous rhythm is mastered perfectly again. Since the synaptic strength w is influenced by the entire learning process, however, w has different coefficients from before.

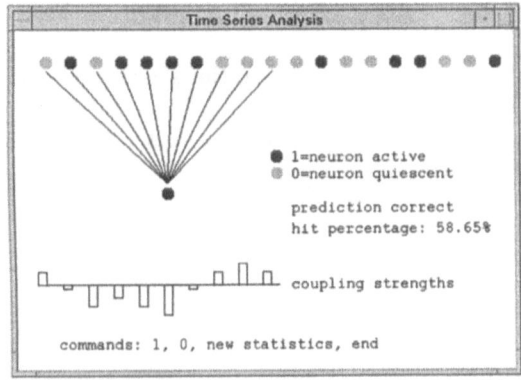

Fig. 3.17. Neural network for
the prediction of a time series

Now we attempt to type in 1 and 0 randomly. We have tested this with students in our class, each of whom wrote about 1000 copies of the two characters to a file, keeping the sequence as random as possible. The evaluation, using nnf, yielded an average hit frequency $g = 60\%$, with values in the range $50\% \leq g \leq 80\%$. Obviously, it is rather difficult to generate random sequences by hand. Frequently, just the prediction $1 \rightarrow 0$ and vice versa yields a hit frequency $g > 50\%$. The perceptron recognizes such rhythms rather quickly.

Thus, five lines of a computer program are capable of predicting our reactions with a certain probability g. On the other hand, we should be able to outwit the network, as the perceptron is a simple deterministic system that we can calculate exactly. In principle, we can therefore calculate which prediction the network will make and then type in precisely the opposite. This should enable us to push the hit frequency g well below 50%.

In fact, one student handed in a sequence that yielded $g = 0$. Later he conceded having known the program before, so he was able to have the network itself generate the opposite bit. Our own attempts at reacting appropriately to the program's predictions were able to suppress g to about 40%. We cordially invite our readers to attempt both: generating random bits and outwitting the perceptron.

Exercise

Manually write a sequence of about 2000 characters to a file name.dat, namely 1 or 0 in as random an order as possible. Extraneous characters are ignored by the program. Get these data evaluated via nnf name.dat. Are you a good random number generator? Check if random bits generated by the computer yield a hit frequency of about 50%. How much is the deviation from 50% allowed to be?

Literature

Hertz J., Krogh A., Palmer R.G. (1991) Introduction to the Theory of Neural Computation. Addison-Wesley, Reading, MA

Müller B., Reinhardt J., Strickland M.T. (1995) Neural Networks: An Introduction. Springer, Berlin, Heidelberg, New York

4. Differential Equations

For more than 300 years, physicists have described the motion of masses by rules in the infinitely small: Under the influence of forces, mass elements move by a minute distance in a minute amount of time. Such a description leads to differential equations which to this day are the most important tool of physics. If twice the cause leads to twice the effect, then the equations can usually be solved. In the case of nonlinear differential equations, however, one is often restricted to few solvable special cases or has to resort to numerical solutions. In this chapter, we want to provide an introduction to some of the numerical methods and, in so doing, solve some simple physics examples.

There are many ready-made program packages to numerically solve differential equations. For example in *Mathematica*, one can use NDSolve to solve ordinary differential equations without knowing the details of the program. Yet we do not want to completely leave out the description of the algorithms, for two reasons: first because for large and difficult problems, it is always necessary to find one's own compromise between the precision of a calculation and the processing time needed, and second because for many special cases there are special methods, which are still being researched today.

4.1 The Runge–Kutta Method

Fundamentals

First, we want to discuss the solution of an ordinary differential equation. To this end, we consider a function $y(x)$ whose nth derivative $y^{(n)}(x)$ can be written as a function of x and all lower derivatives,

$$y^{(n)} = f\left(x, y, y', y'', \ldots, y^{(n-1)}\right) , \tag{4.1}$$

i.e., an explicit nth-order differential equation. This equation is equivalent to n first-order equations, since with $y_1 = y$, $y_2 = y'$, $\ldots$, $y_n = y^{(n-1)}$ one obtains the following set of equations:

$$y_1' = y_2$$
$$y_2' = y_3$$
$$\vdots$$
$$y_{n-1}' = y_n$$
$$y_n' = f(x, y_1, \ldots, y_n) \ ,$$

(4.2)

or, in vector notation using $\boldsymbol{y} = (y_1, y_2, \ldots, y_n)$,

$$\boldsymbol{y}'(x) = \lim_{\Delta x \to 0} \frac{\boldsymbol{y}(x + \Delta x) - \boldsymbol{y}(x)}{\Delta x} = \boldsymbol{f}(x, \boldsymbol{y}) \ . \tag{4.3}$$

If the vector $\boldsymbol{y}$ is known at the point x, then its value after a minute step $x + \Delta x$ can be calculated according to (4.3). This is also the essence of the numerical method: The spatial coordinate x is discretized and an attempt is made at calculating $\boldsymbol{y}$ at the adjacent point $x + \Delta x$ as precisely as possible, using (4.3).

For the description of the numerical method we limit ourselves to the dimension $n = 1$, i.e., $\boldsymbol{y}$ and $\boldsymbol{f}$ each have just one component y and f. The algorithm to be presented can easily be extended to arbitrary dimensions later. So, let us assume

$$y' = f(x, y) \ . \tag{4.4}$$

To start the calculation, we obviously need an initial value $y(x_0) = y_0$. First, we discretize the x-axis:

$$x_n = x_0 + nh \ , \tag{4.5}$$

where n is an integer and h is the step size. Let $y_n = y(x_n)$ and $y_n' = y'(x_n)$. Then the Taylor expansion yields

$$y_{n\pm1} = y(x_n \pm h) = y_n \pm h y_n' + \frac{h^2}{2} y_n'' \pm \frac{h^3}{6} y_n''' + \mathcal{O}(h^4) \ . \tag{4.6}$$

We obtain y_n' from (4.4). In the simplest approximation, the so-called *Euler method*, one thus obtains

$$y_{n+1} = y_n + h f(x_n, y_n) \ . \tag{4.7}$$

The error is of the order h^2. This method is not recommended for practical use. With just a little effort it can be improved significantly.

Equation (4.7) only takes into account the slope f at the point (x_n, y_n), even though that slope will already have changed en route to the next point (x_{n+1}, y_{n+1}). The Taylor series (4.6) shows that it is advantageous to use the slope midway between the two points, as one gains one order in the step size h this way. But to this end one first has to find a suitable estimate for $y_{n+1/2} = y(x_n + h/2)$. This is the idea of the Runge–Kutta method which was developed as early as 1895. If the expansion (4.6) is evaluated at the point $x_n + h/2$, with a step size of $h/2$, then

$$y_n = y_{n+1/2} - \frac{h}{2}y'_{n+1/2} + \left(\frac{h}{2}\right)^2 \frac{1}{2}y''_{n+1/2} + \mathcal{O}\left(h^3\right) \, ,$$

$$y_{n+1} = y_{n+1/2} + \frac{h}{2}y'_{n+1/2} + \left(\frac{h}{2}\right)^2 \frac{1}{2}y''_{n+1/2} + \mathcal{O}\left(h^3\right) \, . \tag{4.8}$$

From this we conclude:

$$y_{n+1} = y_n + hy'_{n+1/2} + \mathcal{O}\left(h^3\right) \, . \tag{4.9}$$

Next, we approximate the derivative $y'_{n+1/2} = f\left(x_{n+1/2}, y_{n+1/2}\right)$ by the term $f\left(x_{n+1/2}, y_n + h/2 f(x_n, y_n)\right) + \mathcal{O}(h^2)$, which leads us to the following algorithm:

$$k_1 = hf\left(x_n, y_n\right) \, ,$$

$$k_2 = hf\left(x_n + \frac{h}{2}, y_n + \frac{k_1}{2}\right) \, ,$$

$$y_{n+1} = y_n + k_2 \, . \tag{4.10}$$

Consequently, this second-order Runge–Kutta method yields an error of order h^3 for only two calculations of the function f. By iterating this step of the calculation several times one can cancel even higher orders of h. The method that has prevailed in practice needs four calculations to achieve an error of order h^5. We only state the result:

$$k_1 = hf\left(x_n, y_n\right) \, ,$$

$$k_2 = hf\left(x_n + \frac{h}{2}, y_n + \frac{k_1}{2}\right) \, ,$$

$$k_3 = hf\left(x_n + \frac{h}{2}, y_n + \frac{k_2}{2}\right) \, ,$$

$$k_4 = hf\left(x_n + h, y_n + k_3\right) \, ,$$

$$y_{n+1} = y_n + \frac{k_1}{6} + \frac{k_2}{3} + \frac{k_3}{3} + \frac{k_4}{6} \, . \tag{4.11}$$

We are left with the problem of choosing the step size h. One possible way of finding an appropriate value for h is to check the result for different values. If the change in y is small, one can afford large step sizes, while many small steps are needed in the opposite case. Thus, if the slope of y varies strongly, the program should reduce the step size h on its own.

To estimate the error, one calculates one Runge–Kutta step (4.11) with a step size $2h$ (result $y_{1\times2}$) in parallel with two steps of step size h (result $y_{2\times1}$). The additional effort this takes is less than 50%, since the derivative at the starting point has already been calculated anyway. If we denote the exact solution for a step from x to $x+2h$ by $y(x+2h)$, then the deviations of the approximate values $y_{2\times 1}$ and $y_{1\times 2}$ respectively from $y(x + 2h)$ are given by

$$y\left(x + 2h\right) = y_{2\times 1} + 2h^5\phi + \mathcal{O}\left(h^6\right) ,$$
$$y\left(x + 2h\right) = y_{1\times 2} + \left(2h\right)^5\phi + \mathcal{O}\left(h^6\right) , \tag{4.12}$$

since we are using a fourth-order method. The constant ϕ is determined by the fifth derivative of the function $y(x)$. For the difference $\Delta = y_{1\times 2} - y_{2\times 1}$, this leads to

$$\Delta \propto h^5 . \tag{4.13}$$

Therefore, if we choose two step sizes h_1 and h_0, we have

$$h_0 = h_1 \left|\frac{\Delta_0}{\Delta_1}\right|^{0.2} . \tag{4.14}$$

We can use this to calculate the step size h_0 required for the desired accuracy Δ_0 from the error Δ_1 calculated for the step size h_1. If the desired accuracy refers to the global rather than the local error used in (4.12) to (4.14), the exponent 1/5 in (4.14) has to be replaced by 1/4. In practice, a mixed procedure has proven reliable, which conservatively uses the exponent 1/5 to calculate the allowed increase of h while using 1/4 for a necessary reduction.

Frequently in physics, equations of motion take the form

$$\dot{x} = v ,$$
$$\dot{v} = f\left(x, t\right) . \tag{4.15}$$

Here, $v(t)$ is the velocity of a particle with the trajectory $x(t)$. In this case, there is a trick for estimating the slope between two points: x is calculated at times $t_n = t_0 + nh$, $n = 1, 2, 3, \ldots$ and v for the times $t_{n+1/2}$ in between by

$$v_{n+1/2} = v_{n-1/2} + hf\left(x_n, t_n\right) ,$$
$$x_{n+1} = x_n + hv_{n+1/2} . \tag{4.16}$$

This means that the algorithm jumps back and forth between calculating the position x and the velocity v in between. This is aptly called the *leapfrog* method. Because the derivatives are always calculated halfway between two consecutive values, the error is of the order h^3. Though the leapfrog method requires the same computation effort as the Euler step, it is more precise and more robust in practice. It is therefore a favorite tool of *molecular dynamics*, i.e., for solving the equations of motion of a system of many interacting particles.

Finally, we want to point out that while the Runge–Kutta method is simple and robust there may be better methods requiring less processing time, depending on the problem. For example, one can calculate the result of multiple simple steps for various values of h and do a suitable extrapolation for $h \to 0$; this is the idea behind the Richardson extrapolation and the

Bulirsch–Stoer method. Another possibility is to extrapolate several y_i values by a polynomial to obtain a prediction for the value of y_n, which is used in turn to calculate the slope y'_n. The value of y_n is then corrected by integrating the values of the y'_i. Suitable algorithms for these so-called predictor–corrector methods can be found in textbooks.

Algorithm

Equation (4.11) is easy to program, even if y and f are replaced again by $\mathbf{y}$ and $\mathbf{f}$. In *Mathematica*, $\mathbf{y}$ and $\mathbf{f}$ are lists of numbers and functions respectively in this case. If $\mathbf{f}$ depends only on $\mathbf{y}$ and not on x, then (4.11) can be written as

```
RKStep[f_, y_, yp_, h_]:=
    Module[{k1, k2, k3, k4},
           k1 = h N[f/.Thread[y -> yp]];
           k2 = h N[f/.Thread[y -> yp+k1/2]];
           k3 = h N[f/.Thread[y -> yp+k2/2]];
           k4 = h N[f/.Thread[y -> yp+k3]];
           yp + (k1 + 2*k2 + 2*k3 + k4)/6 ]
```

The function `Thread` assigns the corresponding value from the list yp to each element of the list y. Note that k1, k2, k3, and k4 are themselves lists. The y_i values in the interval $[0, x]$ can then be calculated from the initial value y_0 via

```
RungeK[f_List,y_List,y0_List,{x_, dx_}]:=
  NestList[RKStep[f,y,#,N[dx]]&,N[y0],Round[N[x/dx]]]
```

For comparison, we also program the Euler method:

```
EulerStep[f_,y_,yp_,h_]:= yp+h N[f/.Thread[y -> yp]]
Euler[f_,y_,y0_,{x_, dx_}]:=
NestList[EulerStep[f,y,#,N[dx]]&,N[y0],Round[N[x/dx]]]
```

Results

We want to solve the differential equation of a problem that we have already investigated extensively in Sect. 1.2: the mathematical pendulum. Its Hamiltonian is

```
hamilton = p^2/2 - Cos[q]
```

Here p is the angular velocity $\dot{\varphi}$ and q is the displacement angle φ of the pendulum. As in Sect. 1.2, energy and time are measured in units of mgl and $\sqrt{l/g}$ respectively. The equations of motion are obtained from the partial derivatives of the Hamiltonian. In *Mathematica*, their solution is then given by

```
RungeK[{D[hamilton,p], -D[hamilton,q]},
              {q,p}, {phi0,p0}, {tmax,dt}]
```

Figure 4.1 shows the result. The curve on the left is the correct result for an initial angle $\varphi_0 = \pi/2$. Without friction, the energy is conserved and the curve must be closed (see Fig. 1.4). For comparison, the plot on the right shows the result of the Euler method using the same step size $h = 0.1$. It can be seen that this method gives a completely wrong result, since the energy increases and the curve is not closed.

Of course, we can easily introduce friction. We add a term of the form $-\mu p$ to the gravitational force and obtain a solution of the equations of motion by

```
RungeK[{p,-Sin[q]-r p},{q,p},{phi0,p0},{tmax,dt}]
```

In Fig. 4.2 we see that the pendulum loses its energy and moves towards its rest position $q = p = 0$.

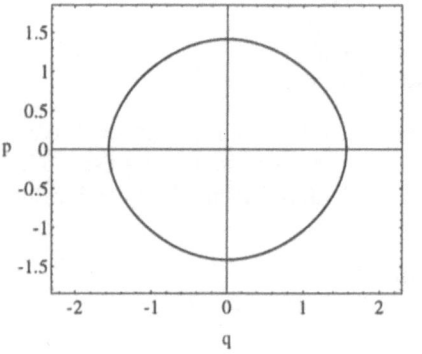 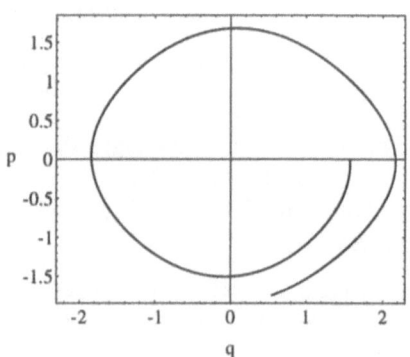

Fig. 4.1. Oscillation of the pendulum in phase space $q = \varphi, p = \dot{\varphi}$. Result of the Runge–Kutta method (*left*) and solution of the same problem calculated with the Euler method (*right*), which obviously gives a wrong result

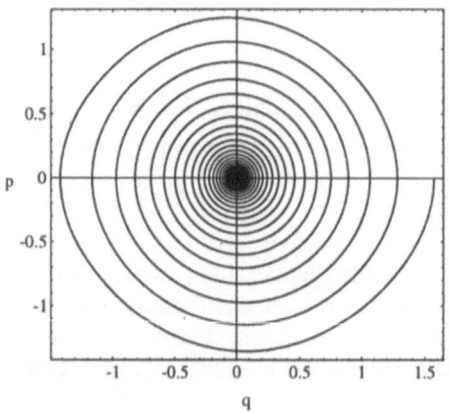

Fig. 4.2. The same oscillation as shown on the left-hand side of Fig. 4.1, but with friction $\mu = 0.05$

Exercise

A particle of mass m moves in a one-dimensional double-well potential and is subject to an additional frictional force proportional to its velocity. Thus the equation of motion for the position $x(t)$ as a function of time t is

$$m\ddot{x} = \mu\dot{x} + ax - bx^3$$

with positive constants μ, a, and b. Depending on the initial state, the particle will end up either in the left $(x < 0)$ or the right potential well.

Using the constants $\mu/m = 0.1$, $a/m = b/m = 1$ (in suitable units), calculate and plot those regions in the plane of initial states $(x(0), \dot{x}(0))$ for which the particle ends up in the left well, i.e., at $x(\infty) = -1$.

Literature

Crandall R.E. (1991) Mathematica for the Sciences. Addison-Wesley, Redwood City, CA

Koonin S.E., Meredith D.C. (1990) Computational Physics, Fortran Version. Addison-Wesley, Reading, MA

Jodl H.-J., Korsch H.J. (1994) Chaos: A Program Collection for the PC. Springer, Berlin, Heidelberg, New York

Press W.H., Teukolsky S.A., Vetterling W.T., Flannery B.P. (1992) Numerical Recipes in C: The Art of Scientific Computing. Cambridge University Press, Cambridge, New York

Schmid E.W., Spitz G., Lösch W. (1990) Theoretical Physics on the Personal Computer. Springer, Berlin, Heidelberg, New York

Stoer J., Bulirsch R. (1996) Introduction to Numerical Analysis. Springer, Berlin, Heidelberg, New York

DeVries P.L. (1994) A First Course in Computational Physics. Wiley, New York

4.2 The Chaotic Pendulum

The French mathematician Henri Poincaré showed, as early as the end of the nineteenth century, that a simple mechanical system can exhibit complex behavior. The idea that one merely has to specify the initial conditions as accurately as possible to be able, in principle, to precisely predict the motion of a mechanical system from its Newtonian equations of motion is, in practice, taken to the point of absurdity, even by simple models. In general, a system reacts so sensitively to the initial conditions that a minute uncertainty of the initial values will lead to a large uncertainty in the prediction after just a short time. This is true not only for the weather forecast, but also for a simple, externally driven pendulum.

Though this fact has been known for a long time, it took almost 100 years for its significance to be recognized and investigated by the scientific community. Only the computer made it feasible to investigate chaotic motion in detail and – what is possibly just as important – display it graphically.

In this section, we want to numerically calculate a simple example: the nonlinear pendulum with frictional force, which is driven by a periodic external force.

Physics

We first consider the pendulum from Sect. 1.2 and add a frictional force. In dimensionless form, the equation of motion for the displacement angle $\varphi(t)$ is

$$\ddot{\varphi} + r\dot{\varphi} + \sin\varphi = 0 \tag{4.17}$$

with a friction coefficient $r \geq 0$. This second-order differential equation can be rewritten in the form of two first-order equations by introducing $\omega(t) = \dot{\varphi}(t)$:

$$\dot{\varphi} = \omega \,,$$
$$\dot{\omega} = -r\omega - \sin\varphi \,. \tag{4.18}$$

The motion of the pendulum can therefore be represented in two-dimensional phase space. Each value (φ, ω) uniquely determines its change $(d\varphi, d\omega)$ over a time interval dt. Therefore, an orbit $(\varphi(t), \omega(t))$ cannot intersect itself.

It can be shown that this means that there cannot be any chaotic motion. In two dimensions, there is simply no room for trajectories that do anything other than form closed orbits or approach them. In our case, the pendulum relaxes to the rest position ($\varphi = 0, \omega = 0$) for $r > 0$, possibly after several oscillations or full turns. The origin of phase space is an attractor for nearly all starting points $(\varphi(0), \omega(0))$. Eventually, all orbits spiral towards the rest position (see Fig. 4.2).

The picture changes if we allow a third direction in phase space. To do so, we drive the pendulum by a periodic torque of strength a and frequency ω_D,

$$\ddot{\varphi} + r\dot{\varphi} + \sin\varphi = a\cos(\omega_D t) \,. \tag{4.19}$$

Using $\theta = \omega_D t$ we obtain

$$\dot{\varphi} = \omega \,,$$
$$\dot{\omega} = -r\omega - \sin\varphi + a\cos\theta \,,$$
$$\dot{\theta} = \omega_D \,. \tag{4.20}$$

The motion of the pendulum is now described in three-dimensional space $(\varphi, \omega, \theta)$. There are three parameters (r, a, ω_D). Without driving torque and friction ($a = r = 0$) and for small angles $\varphi \ll 1$ we get a harmonic oscillation with frequency $\omega_0 = 1$. Therefore there are, generally, three competing time scales: the period $2\pi/\omega_D$ of the driving torque, the period $2\pi/\omega_0$ of the free oscillation, and the relaxation time $1/r$. Here, just as before in the cases of the Hofstadter butterfly and the Frenkel–Kontorova model, the competition between different length or time scales leads to interesting physical phenomena.

The three-dimensional motion $(\varphi(t), \omega(t), \theta(t))$ is difficult to analyze, even in cases like ours, where θ is just a linear function of time. In order to reduce the multitude of possible forms of motion to the essential structures, the motion is observed only at fixed time intervals, as if the system were illuminated stroboscopically. The result is called a *Poincaré section*.

For the time interval, we take the period of the driving torque and consider

$$(\varphi(t_j), \omega(t_j)) \quad \text{with} \quad t_j = \frac{2\pi j}{\omega_D} \quad \text{and} \quad j = 0, 1, 2, \ldots . \qquad (4.21)$$

This gives us a sequence of points in the plane, where each point is uniquely determined by its predecessors. As in the case of the chain on a corrugated surface (Sect. 3.2), we reduce the problem to a two-dimensional discrete function, which this time, however, is not area-preserving, if $r > 0$. Rather, an area segment gets smaller in the iteration and attractors are generated, as for the one-dimensional logistic map (Sect. 3.1).

What do the orbits $(\varphi(t_j), \omega(t_j))$ look like in the Poincaré section? For a periodic motion with period $2\pi/\omega_0$, the stroboscopic exposure reveals either individual points or a closed curve. For $\omega_0 = \omega_D$, we get a single point in the (φ, ω) plane. For $\omega_0 = (p/q)\omega_D$ with p and q integer and relatively prime, one obtains q different points, and p determines the order in which the q points are accessed. If, on the other hand, ω_0 is an irrational multiple of ω_D, we get an infinite number of points $(\varphi(t_j), \omega(t_j))$ filling a closed curve. Thus, periodic motion results in either individual points or closed curves in the Poincaré section. Such orbits can be attractors, i.e., if one starts with points $(\varphi(0), \omega(0))$ in a certain basin of attraction, then all these values move towards such a periodic attractor. Multiple attractors with their associated basins of attraction are possible as well.

There is a second kind of attractors, though, so-called *strange attractors*. They can be visualized as a kind of puff pastry dough, which is obtained by repeated stretching and folding. Such attractors correspond to chaotic orbits which traverse phase space in a seemingly unpredictable manner. The strange attractors are fractals (see Sect. 3.3), i.e., they are more than a line, but less than an area.

As in Sect. 3.3, the fractal dimension D can be determined by covering the attractor with N squares whose edges have the length ε:

$$D_B = - \lim_{\varepsilon \to 0} \frac{\log N(\varepsilon)}{\log \varepsilon} . \qquad (4.22)$$

A method that is numerically more efficient was suggested by Grassberger and Procaccia: We generate N points x_i on the attractor, which should have as little correlation as possible. Then we count the number of points x_j whose distance from x_i is less than a given value R, and average this number over the x_i. We can formally express this correlation using the step function $\Theta(x)$:

$$C(R) = \lim_{N \to \infty} \frac{1}{N(N-1)} \sum_{\substack{i \neq j}}^{N} \Theta(R - |x_i - x_j|) \ . \tag{4.23}$$

$C(R)$ can be interpreted as the average mass of a section of the attractor, and the relation between mass and length already used in Sect. 3.3 defines a fractal dimension D_C via

$$C(R) \propto R^{D_C} \ . \tag{4.24}$$

A log–log plot of this equation should therefore yield a straight line with slope D_C. This, however, is only true for values of R which are larger than the average distance between the data points, and smaller than the size of the attractor.

Another method for determining the fractal dimension uses the concept of information entropy. Again, the attractor is covered with squares whose edges have the length ε. Then one counts how many points of the orbit generated lie in each square. Let p_i be the probability that a point (φ, ω) of the attractor is found in square i. Then the entropy is defined as

$$I(\varepsilon) = - \sum_i p_i \ln p_i \ , \tag{4.25}$$

and the information dimension D_I is obtained from

$$D_I = - \lim_{\varepsilon \to 0} \frac{I(\varepsilon)}{\ln \varepsilon} \ . \tag{4.26}$$

Now it can be shown that the following relation between the three dimensions holds:

$$D_B \leq D_I \leq D_C \ . \tag{4.27}$$

In practice the three dimensions often agree within the experimental error.

Algorithm

To solve the differential equation (4.20) numerically, we choose the fourth-order Runge–Kutta method from the previous section. As we want to be able to display the movement on the screen while the calculation is still in progress, we use the language C with the appropriate graphics environment. In our program the strength a of the driving torque can be changed by a keystroke during the run; also, we can switch back and forth between displaying the continuous motion and the Poincaré section.

Programming the Runge–Kutta step from (4.11) oneself is not difficult, but we want to demonstrate at this point how to incorporate the routine **odeint** from *Numerical Recipes* into one's own program.

The function odeint integrates the system of equations (4.20) using an adaptive control of the step size. Therefore the points of the orbit that are calculated do not represent its actual development in time. In order to observe that correlation, one should directly use the routine rk4 from *Numerical Recipes*.

First, all variables of the routines used must be declared, and the routines must be added to one's own program. In doing so, it has to be noted that odeint uses the programs rk4 and rkqc and that all routines call certain error-handling routines which are contained in nrutil. One can directly copy all programs into one's own code or add them via #include by, e.g.,

```
#define float double
#include "\tc\recipes\nr.h"
#include "\tc\recipes\nrutil.h"
#include "\tc\recipes\nrutil.c"
#include "\tc\recipes\odeint.c"
#include "\tc\recipes\rkqc.c"
#include "\tc\recipes\rk4.c"
```

If you only have access to *Numerical Recipes 2.0*, you should replace rkqc.c with rkqs.c and rk4.c with rkck.c. The path name has to be modified, of course, to indicate where you have stored the programs. We have added the first command so that all real variables in the program and in the *Numerical Recipes* are of the same type double. The routine odeint is called in the following form:

```
odeint(y,n,t1,t2,eps,dt,0.,&nok,&nbad,derivs,rkqc)
```

Here, y is a vector with the components $(\varphi(t_1), \omega(t_1))$. It is integrated over the interval $[t_1, t_2]$ and then y is replaced by the result $(\varphi(t_2), \omega(t_2))$. The variable n denotes the number of variables. Since we replace θ by $\omega_\mathrm{D} t$ again in (4.20), there are only n = 2 variables. With eps we can specify the desired accuracy, and dt is an estimate of the step size required. On output, the variables nok and nbad contain information about the number of steps needed. The parameter derivs is the name of a function that is used to evaluate the right-hand side of (4.20). In other words, derivs(t,y,f) calculates the components of f from t and y. In our example, we have

$$y[1] = \varphi \,,$$
$$y[2] = \omega \,,$$

and correspondingly

$$f[1] = y[2] \,,$$
$$f[2] = -\,r * y[2] \;-\; \sin(y[1]) \;+\; a * \cos(\omega_\mathrm{D} * t) \,.$$

Finally, the routine rkqc, which in turn calls the Runge–Kutta step rk4, is passed to odeint. One can, however, pass other integration procedures, possibly one's own, just as well.

In summary, then, the essential part of the program **pendulum**.c consists of the following commands:

```
main()
{
y[1]=pi/2.;
y[2]=0.;

while(done==0)
 {
   odeint(y,2,t,t+3.*pi,eps,dt,0.,&nok,&nbad,derivs,rkqc);
   xold=fmod(y[1]/2./pi +100.5,1.)*ximage;
   yold=y[2]/ysc*yimage/2+yimage/2;
   rectangle(xold,yold,xold+1,yold+1);
   t=t+3.*pi;
 }
}

void derivs(double t,double *y,double *f)
 {
   f[1]=y[2];
   f[2]=-r*y[2]-sin(y[1])+a*cos(2./3.*t);
 }
```

Here, too, a good programmer should have all constants calculated in advance. The vectors y and f were declared longer than necessary, by one element, because the indices in the *Numerical Recipes* start at 1 rather than 0 (as is customary in C).

The program above uses $\omega_D = 2/3$ and plots the points $(\varphi(t_i), \omega(t_i))$ (as small rectangles), each of them one period $[t, t + 3\pi]$ of the driving torque after the previous one. In order to observe the continuous orbit, one has to replace $t + 3\pi$ by $t + dt$, where the time step dt should be chosen such that the orbit progresses by only a few pixels each time step. In this case, of course, a line from the initial to the final point is plotted instead of a single point.

Results

As an example we investigate on screen the forced pendulum with a friction coefficient $r = 0.25$ and a driving frequency $\omega_D = 2/3$. For the initial angle we pick $\varphi(0) = \pi/2$ and $\omega(0) = 0$ for the initial angular velocity. The keys i and d respectively allow us to increment and decrement the driving torque a in steps of 0.01.

Without external torque ($a = 0$), the pendulum executes a damped oscillation and eventually stops at the rest position $(\varphi, \omega) = (0,0)$. For small values of a we obtain, after a transient phase, a periodic motion with a period $2\pi/\omega_D$, i.e., a single point in the Poincaré section. As a reaches a value of 0.68, the drive is strong enough that the pendulum can turn over at the top. The orbit leaves the screen and reenters on the opposite side, owing to the periodicity of the angle φ (=horizontal axis).

For $a = 0.7$, we observe a chaotic attractor (Fig. 4.3). In the Poincaré section, a fractal "puff pastry dough" begins to show up (Fig. 4.4). For $a = 0.85$, the orbit still looks irregular (Fig. 4.5), but the Poincaré section clearly shows a cycle of length $7 \times 2\pi/\omega_\mathrm{D}$. Even period doubling can be observed. The single cycle, with a turnover at $a = 0.97$, doubles for $a = 0.98$ and $a = 0.99$, and quadruples at $a = 1.00$. By the time a reaches 1.01, the orbit looks chaotic yet again. For $a = 1.2$, three points can be seen, and for $a = 1.3$, we get a chaotic orbit with three bands, whose Poincaré section is shown in Fig. 4.6.

As the strength a of the drive increases, we observe a continuous change back and forth between periodic and chaotic motion.

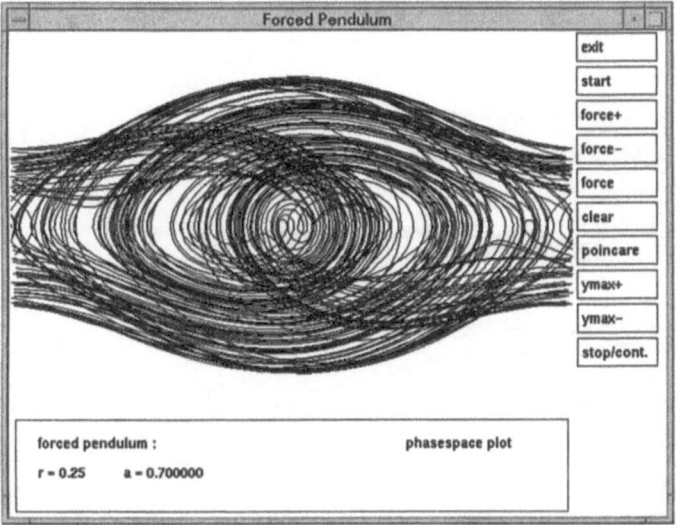

Fig. 4.3. Motion of the forced pendulum in (φ, ω) phase space; the driving torque a is 0.7

Exercise

A liquid of given viscosity μ, heat conductivity κ, and density ρ is enclosed between two plates. The upper plate has a temperature T, the lower a temperature $T + \Delta T$. If the temperature difference is small, only heat conduction takes place; any convective motion is prevented by viscous resistance. As the temperature difference increases, the hot liquid starts to rise in some areas and sink again in others; so-called convection cells result. As the temperature increases even further, these static cells are broken up and chaotic motion results.

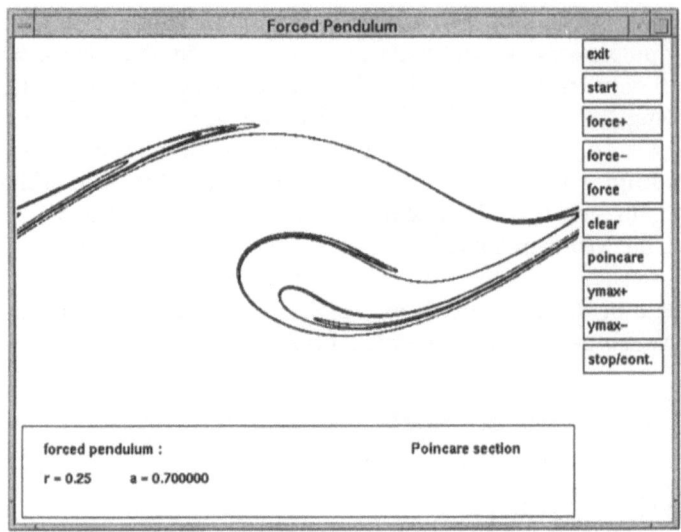

Fig. 4.4. Poincaré section for the motion from Fig. 4.3, taken at time intervals $2\pi j/\omega_D$

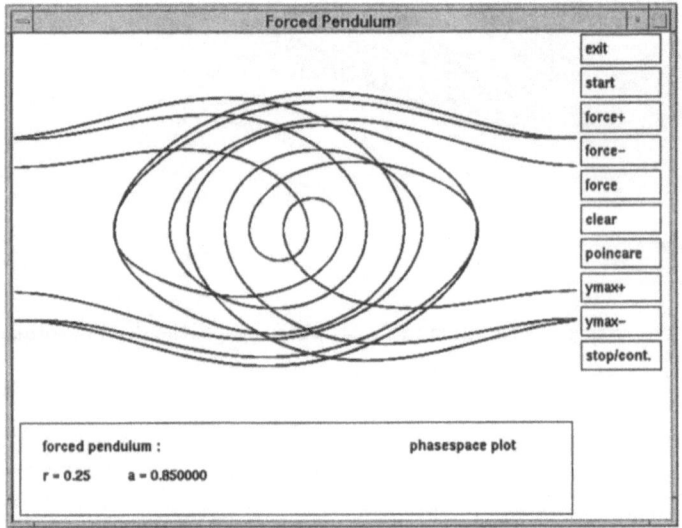

Fig. 4.5. Phase-space plot for $a = 0.85$. The motion is a cycle of period $7 \times 2\pi/\omega_D$. The corresponding Poincaré section shows exactly seven points

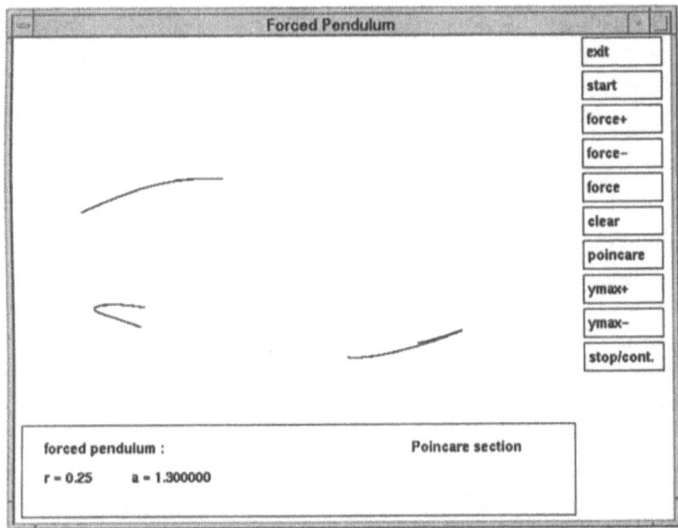

Fig. 4.6. Poincaré section for $a = 1.3$. The motion alternates between three chaotic orbits

Assuming that the convection cells extend to infinity in the y-direction, the meteorologist Lorenz has developed a model that gives a good description of such a Bénard experiment near the transition from ordered to chaotic behavior. The model uses a Fourier expansion in the x- and z-directions and neglects higher-order terms of the Fourier series.

In this approximation the system is described by the following three equations:

$$\dot{X} = -\sigma X + \sigma Y ,$$
$$\dot{Y} = -XZ + rX - Y ,$$
$$\dot{Z} = XY - bZ .$$

Here, X represents the velocity of the circular motion, Y the temperature difference between the rising and the sinking liquid, and Z represents the deviation of the resulting temperature profile from the equilibrium profile. The parameters b and σ are determined solely by the material constants (μ, ρ, κ) and the geometric dimensions; consequently, they have to be regarded as constants. The quantity r is proportional to the temperature difference applied and thus serves as an external control parameter that determines the system's behavior.

This system is interesting for various applications:

- in meteorology: air movement;
- in astronomy: stellar structure of convective stars;
- in energy technology: heat conduction of insulating materials.

Integrate the differential equations of the Lorenz model with the Runge–Kutta method. Use $\sigma = 10$ and $b = 8/3$. The parameter r should be interactively controlled by the user. Graphically display the result for X and Y, specifically

- as a projection of the continuous motion onto the X–Y plane,
- as a Poincaré section for $Z = \text{constant} = 20.0$.

To better determine the intersection with the plane $Z = \text{constant} = 20$, a linear interpolation between the last value below 20 and the first one above 20 should be used.

When does chaotic behavior set in? For $r = 28$, what is the dimension of the strange attractor in the Poincaré section mentioned above?

Literature

Baker G.L., Gollub J.P. (1996) Chaotic Dynamics: An Introduction. Cambridge University Press, Cambridge, New York

Jodl H.-J., Korsch H.J. (1994) Chaos: A Program Collection for the PC. Springer, Berlin, Heidelberg, New York

Ott E. (1993) Chaos in Dynamical Systems. Cambridge University Press, Cambridge, New York

Press W.H., Teukolsky S.A., Vetterling W.T., Flannery B.P. (1992) Numerical Recipes in C: The Art of Scientific Computing. Cambridge University Press, Cambridge, New York

4.3 Stationary States

Newton's equations of motion, however successfully they describe our macroscopic world, turn out to be unsuitable if electrons in atoms, molecules, or solids are concerned. In the realm of the microscopically small, only probability statements for the position and momentum of a particle are possible; their time dependence is determined by the quantum-mechanical Schrödinger equation. Under certain conditions, this dynamic equation reduces to the so-called stationary Schrödinger equation, an eigenvalue equation for the energy operator, which in the coordinate representation takes the form of a linear differential equation. In principle, this equation can be solved numerically using the methods from the previous sections. There is, however, a more efficient method for numerically solving the Schrödinger equation, which we want to illustrate using a simple example, the anharmonic oscillator. For the Schrödinger equation we have the additional problem that the energy cannot take all real values, but only certain discrete ones, which are determined by the normalizability requirement.

Physics

A quantum particle is described by a complex-valued (in general) wave function $\psi(r,t)$ that obeys a second-order partial differential equation. $|\psi(r,t)|^2$ is the probability density for finding the particle at the position r at time t.

If the mean values of all measurable quantities in this state are independent of time, the wave function ψ obeys the stationary Schrödinger equation, which, in one spatial dimension, has the following form:

$$-\frac{\hbar^2}{2m}\psi''(x) + V(x)\,\psi(x) = E\psi(x) \ . \tag{4.28}$$

Since the potential $V(x)$ is real, we can choose ψ as a real-valued function too. Equation (4.28) describes, for one dimension, the stationary state of a particle of mass m in the potential $V(x)$; the equation is linear and contains a second derivative. Compared to Newton's equation of motion, there is one important difference, though: the energy E cannot take arbitrary values; instead (4.28), combined with the requirement that the wave function has to be normalizable, i.e., that $\int dx |\psi(x)|^2$ has to have a finite value, determines the possible values of E. It turns out that even a tiny deviation of E from an allowed energy value causes the wave function to grow exponentially for large x. This fact has to be taken into account in the numerical search for a solution. As we will see, on the other hand, one can take advantage of exactly this behavior to numerically determine the energy eigenvalues very precisely.

For symmetric potentials $V(x) = V(-x)$, the stationary state, too, has the symmetry $\psi(x) = \pm\psi(-x)$, where the sign alternates with increasing energy values. The ground state is symmetric, $\psi_0(x) = \psi_0(-x)$, and has no zero; $|\psi_0(x)| > 0$. With each higher energy eigenvalue, the number of zeros (= nodes) of the wave function increases by one. Thus the number of nodes of the wave function $\psi(x)$ specifies the number of the energy level.

In our example, we study the anharmonic oscillator from Sect. 2.1, whose Schrödinger equation has the following dimensionless form:

$$\psi''(x) + \left(-x^2 - \lambda x^4 + 2E\right)\psi(x) = 0 \ . \tag{4.29}$$

Here the energy E and the position x are measured in units of $\hbar\omega$ and $\sqrt{\hbar/(m\omega)}$ respectively. In the harmonic case ($\lambda = 0$), we know the eigenvalues and eigenstates from analytic calculations. Relative to the ground state energy $E_0^0 = 1/2$, E^0 can only take integer nonnegative values,

$$E_n^0 = n + \frac{1}{2} \quad \text{where} \quad n = 0, 1, 2, \dots \ . \tag{4.30}$$

These energy levels are shifted to higher values by an anharmonic term λx^4 with $\lambda > 0$. For negative values of λ, there are no stationary states, since in this case the negative term proportional to x^4 dominates for large $|x|$ and the potential $V(x)$ gets more and more negative. Owing to the tunnel effect any wave function will diffuse towards infinity. This means the $E_n(\lambda)$ cannot be analytic at $\lambda = 0$.

Algorithm

Several problems have to be solved: How can the eigenvalues E be found? Which algorithm does one use to integrate the Schrödinger equation? Which initial values $\psi(x_0)$ and $\psi(x_1)$ are to be used in the numerical integration?

We solve the first problem by using the so-called shooting method: we pick an energy value E for which we know for sure that it is below the ground state energy E_0, e.g., $E = 0.5$, and integrate the Schrödinger equation from the starting point x_0 up to an x value where $|\psi(x)|$ has become unrealistically large. Then we increase the energy step by step, until $\psi(x)$ changes sign. Now, according to the node theorem we have found an energy interval which must contain E_0. Next, we do the integration with an energy value in the middle of the interval; if $\psi(x)$ changes sign, E_0 is in the lower half of the interval, otherwise it is in the upper half. In this way, we can quickly close in on E_0.

One could say that we "shoot" the wave function $\psi(x)$ to large x values for different values of E until the boundary condition $|\psi(x)| \to 0$ for large x values is approximately fulfilled. However, $\psi(x)$ only stays close to zero in a certain interval before exploding towards infinity again.

Even though the second problem, integrating the Schrödinger equation, can in principle be solved by using the Runge–Kutta method from Sect. 4.1, it is possible to achieve a higher accuracy for this type of equation with a simpler method, the Numerov method. Equation (4.29) is of the form

$$\psi''(x) + k(x)\psi(x) = 0 \tag{4.31}$$

with $k(x) = 2(E - V(x))$. We discretize the x-axis with a step size h, specifically $x_n = (n - 1/2)h$ for even wave functions and $x_n = nh$ for odd ones, and designate the wave function and its derivatives at the points x_n by a subscript n, i.e., $\psi_n = \psi(x_n)$, $\psi_n' = \psi'(x_n)$, and so on for the higher derivatives. From the Taylor expansion,

$$\psi_{n\pm1} = \psi_n \pm h\psi_n' + \frac{h^2}{2}\psi_n'' \pm \frac{h^3}{6}\psi_n^{(3)} + \frac{h^4}{24}\psi_n^{(4)} \pm \dots , \tag{4.32}$$

we obtain

$$\psi_{n+1} + \psi_{n-1} = 2\psi_n + h^2\psi_n'' + \frac{h^4}{12}\psi_n^{(4)} + \mathcal{O}\left(h^6\right) . \tag{4.33}$$

For ψ_n'' we can substitute

$$\psi_n'' = -k_n \psi_n , \tag{4.34}$$

according to (4.31). Now we get to the decisive trick, replacing the fourth derivative by the second one in its discretized form:

$$\begin{aligned}
\psi_n^{(4)} &= \left.\frac{d^2}{dx^2}\psi''(x)\right|_{x_n} = \left.-\frac{d^2}{dx^2}(k(x)\psi(x))\right|_{x_n} \\
&= -\frac{k_{n+1}\psi_{n+1} - 2k_n\psi_n + k_{n-1}\psi_{n-1}}{h^2} + \mathcal{O}\left(h^2\right) .
\end{aligned} \tag{4.35}$$

By inserting this into (4.33) we obtain

$$\left(1 + \frac{h^2}{12} k_{n+1}\right) \psi_{n+1} - 2\left(1 - \frac{5}{12} h^2 k_n\right) \psi_n$$

$$+ \left(1 + \frac{h^2}{12} k_{n-1}\right) \psi_{n-1} = 0 + \mathcal{O}\left(h^6\right) . \qquad (4.36)$$

With this we can calculate the complete set of ψ_n values from two adjacent initial values.

How are the two initial values selected? As we have already mentioned, the eigenfunctions are alternately even and odd as the energy values increase, owing to the symmetry of the potential. Correspondingly, we pick $\psi_0 = \psi_1 \neq 0$ for the even wave functions and $\psi_0 = 0, \psi_1 \neq 0$ for the odd ones. The value of ψ_1 can be selected arbitrarily, as it does not affect the energy E_k, owing to the linearity of the Schrödinger equation, but only the value of the normalization $\int |\psi(x)|^2 dx$.

This defines the algorithm, and we can easily program it. As our programming language we pick C in order to be able to immediately display the wave function on the screen and interactively close in on the energy values E_k via keyboard input.

The integration step (4.36) turns into the function step(&x,dx,&y,&ym1), which yields yp1 ($= \psi_{n+1}$) as its result. Here, x is the current value of x_n, dx is the step size, and y, ym1 correspond to ψ_n and ψ_{n-1}. Since we want to change the values of x, y, ym1 from within step, we have to pass the addresses of these variables, each of which is marked by an &. Note that in C arguments to a function are passed by value and no values are returned to the arguments.

With this, the calculation of ψ_{n+1} goes as follows:

```
double step(double *xa, double dx, double *ya,
            double *ym1a)
{
   long i,n;
   double k(double);
   double yp1, x, y, ym1, xp1, xm1;
   x=*xa; y=*ya; ym1=*ym1a;
   n=ceil(10./dx/500.);
   for(i=1;i<=n;i++)
      {
              xp1=x+dx;   xm1=x-dx;
              yp1=(2.*(1.-5./12.*dx*dx*k(x))*y
                  -(1.+dx*dx/12.*k(xm1))*ym1)/
                   (1.+dx*dx/12.*k(xp1));
              xm1=x;   x=xp1;
              ym1=y;   y=yp1;
      }
   *xa=x;*ya=y;*ym1a=ym1;
   return yp1;
}
```

One can accelerate the calculations that are repeated over and over in this
loop by writing constants like 5./12.*dx*dx to a variable beforehand. The
step size dx covered by the iteration loop for (...) corresponds to exactly
one pixel of the x-axis on the screen; (xs,ys) are the screen coordinates. The
value of xs is increased by another for loop in increments of 1, while ysnew
is calculated from yp1. As an initial value for ψ_1 we select 1.0. With this, the
main loop for the even wave functions becomes

```
x=dx/2.; y=ym1=1.; ysold=1;
for(xs=1; xs<XMAX; xb++)
   {
      yp1=step(&x,dx,&y,&ym1);
      ysnew=(1.-yp1)*YMAX;
      if(abs(ysnew)>10000) break;
      line(xs-1,ysold,xs,ysnew);
      ysold=ysnew;
   }
```

Here, the explosion criterion used for ψ_n is that the y-coordinate exceeds the
value 10000. On the PC a line from (xs-1,ysold) to (xs,ysnew) is drawn.
After this loop, the program waits for a keyboard input via ch=getch() and,
depending on the input, the value of e $(= E)$, the energy step size de, or the
step size dx $(= h)$ of the integration is changed.

Result

The program schroedinger calculates and plots the wave function $\psi(x)$ for
a value $\lambda = 0.1$ of the strength of the anharmonic potential. Figure 4.7
shows the result for $E = 6.22030088$. The symmetry $\psi(x) = \psi(-x)$ (only
the range $x \geq 0$ is plotted) means that the wave function has four zeros,
which means that we are looking for E_4, the fourth energy level above the
ground state. For $\lambda = 0$, that energy is $E_4^0 = 4.5$. Thus, the anharmonic term
with $\lambda = 0.1$ significantly shifts the energy level upwards. The wave function
$\psi(x)$ from Fig. 4.7 is already very close to the bound state $\psi_4(x)$. This can
be seen from the fact that $\psi(x)$ is close to the x-axis for a long time before
shooting exponentially towards $+\infty$. If E is increased by 0.00000005, then
the corresponding $\psi(x)$ diverges towards $-\infty$. Consequently, the energy E_4
falls within the interval

$$[6.22030088, 6.22030093] \ .$$

In Sect. 2.1 we calculated the energy levels by using a different method.
There, the infinite matrix of the Hamiltonian was approximated by an $N \times N$
submatrix and diagonalized. For $N = 50$, we found the value

$$E_4 = 6.220300900006516 \ ,$$

which agrees with the value above. The advantage of the direct integration is
that we get an upper and a lower limit. We can check the accuracy of these
limits by cutting the step size dx of the integration in half.

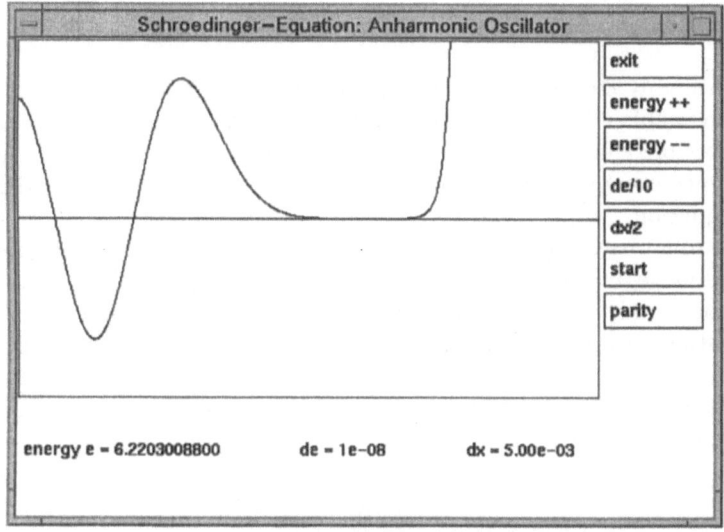

Fig. 4.7. Wave function $\psi_4(x)$ with four nodes for the fifth energy eigenvalue

Exercise

In Sect. 2.1, the four lowest energy levels of the double-well potential were to be calculated by using the matrix method. Here, the same problem is to be solved by integrating the Schrödinger equation. In dimensionless form, the potential V is

$$V(x) = -2x^2 + \frac{x^4}{10}.$$

Using the shooting method, calculate the lowest energies and the associated wave functions and compare your results to those from Sect. 2.1.

Literature

Dahmen H.D., Brandt S. (1994) Quantum Mechanics on the Personal Computer. Springer, Berlin, Heidelberg, New York

Koonin S.E., Meredith D.C. (1990) Computational Physics, Fortran Version. Addison-Wesley, Reading, MA

Schmid E.W., Spitz G., Lösch W. (1990) Theoretical Physics on the Personal Computer. Springer, Berlin, Heidelberg, New York

Stoer J., Bulirsch R. (1996) Introduction to Numerical Analysis. Springer, Berlin, Heidelberg, New York

4.4 Solitons

So far, we have only considered differential equations for functions that depend on *one* variable (time or position). Now we want to study a wave that moves and possibly changes its shape as a function of time. In this case, the time dependence is coupled to the spatial dependence and one gets a partial differential equation.

In this section, we want to deal with the *Korteweg–de Vries equation* (KdV equation) and make an exception by numerically reproducing a solution that can be obtained analytically as well. The KdV equation is a nonlinear differential equation with special solutions, the so-called solitons, that exhibit a number of surprising properties. For example, two solitons can move towards one another, penetrate one another and then move on at their original velocity without changing shape.

Physics

In the theory of water waves in shallow channels, the deviation $u(x,t)$ from the mean water level at time t and position x is described by a nonlinear partial differential equation. In dimensionless and suitably normalized form, the KdV equation can be written as

$$\frac{\partial u}{\partial t} = 6u\frac{\partial u}{\partial x} - \frac{\partial^3 u}{\partial x^3} \ . \tag{4.37}$$

The contribution from the term $u\partial u/\partial x$ increases fourfold if the deviation is doubled, which means that the equation is nonlinear.

Among the many solutions of the KdV equation, there is one that is rather characteristic, namely

$$u\left(x,t\right) = -2\,\mathrm{sech}^2\left(x - 4t\right)\ , \tag{4.38}$$

with $\mathrm{sech}\,x = 1/\cosh x$. This function $u(x,t)$ describes a wave trough that moves to the right at a velocity $v = 4$, not changing its shape in the process. Such a solution is called a soliton.

More complex solutions are known as well, though. The mathematical theory for this is not simple, which is why we only state the result here. The initial state,

$$u\left(x,0\right) = -N\left(N+1\right)\,\mathrm{sech}^2\left(x\right)\ , \tag{4.39}$$

results in N solitons that propagate at different velocities. For example, for $N = 2$, one finds

$$u\left(x,t\right) = -12\frac{3 + 4\cosh\left(2x - 8t\right) + \cosh\left(4x - 64t\right)}{\left[3\cosh\left(x - 28t\right) + \cosh\left(3x - 36t\right)\right]^2} \ . \tag{4.40}$$

Figure 4.8 shows that this solution contains two solitons. A wave with a large amplitude has overtaken a broader wave with a small amplitude. At time

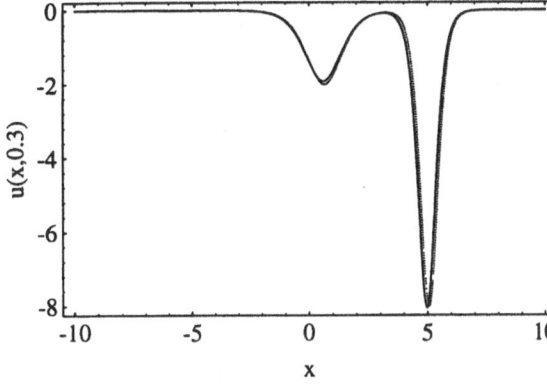

Fig. 4.8. Two solitons move away from one another. The numerical solution (*solid curve*) and the exact one (*dotted curve*) are in good agreement ($t = 0.3$, $\mathrm{d}x = 0.18$, $\mathrm{d}t = 0.002$)

$t = 0$, the two overlap to form the wave packet (4.39), whereas after their encounter both solitons have resumed their original shapes.

We would also like to mention an interesting link to quantum mechanics: The Schrödinger equation with the potential $V(x) = -N(N + 1) \operatorname{sech}^2(x)$ has N bound states, from which one can construct the solutions of the KdV equation by using the methods of inverse scattering theory. Since the explanation is too complicated for the scope of this book, we refer the reader to the literature about solitons.

Algorithm and Results

In order to numerically solve the KdV equation (4.37), we must first discretize the space and time coordinates x and t,

$$u_n^j = u(j\mathrm{d}x, n\mathrm{d}t) \ . \tag{4.41}$$

Here j and n are integers, and $\mathrm{d}x$ and $\mathrm{d}t$ are the step sizes in the x and t coordinates respectively. We want to demonstrate what happens if we write (4.37) too naively as a difference equation. For example, we could consistently use the so-called forward-two-point formula for each derivative $f'(x_k)$, i.e., approximate $f'(x_k)$ by $(f(x_k + h) - f(x_k))/h$. For the partial derivatives this yields

$$\frac{\partial u}{\partial t} \ \rightarrow \ \frac{1}{\mathrm{d}t}\left(u_{n+1}^j - u_n^j\right) \ , \quad \frac{\partial u}{\partial x} \ \rightarrow \ \frac{1}{\mathrm{d}x}\left(u_n^{j+1} - u_n^j\right) \ . \tag{4.42}$$

By applying this rule three times we obtain

$$\frac{\partial^3 u}{\partial x^3} \ \rightarrow \ \frac{1}{(\mathrm{d}x)^3}\left(u_n^{j+3} - 3u_n^{j+2} + 3u_n^{j+1} - u_n^j\right) \ . \tag{4.43}$$

With these substitutions, we solve (4.37) for u_{n+1}^j with the result

$$u_{n+1}^j = u_n^j + \mathrm{d}t\left(6u_n^j\frac{u_n^{j+1} - u_n^j}{\mathrm{d}x} - \frac{u_n^{j+3} - 3u_n^{j+2} + 3u_n^{j+1} - u_n^j}{(\mathrm{d}x)^3}\right) \ . \tag{4.44}$$

This equation is easily programmable. In *Mathematica*, we represent $\{u_n^j\}_{j=0}^{\text{max}}$ by a list with $\text{max} + 1$ elements. To initialize it, we use (4.39) with $N = 2$:

```
ustart:=Table[-6 Sech[(j-max/2)dx]^2//N,{j,0,max}]
```

and assume periodic boundary conditions. We obtain the shifted lists $\{u_n^{j+k}\}$ for $k = 1, 2$, and 3 via `uplus[k]=RotateLeft[u,k]` and can then formulate the integration step $\{u_n^j\} \rightarrow \{u_{n+1}^j\}$ as follows:

```
step[u_]:=(Do[uplus[k]=RotateLeft[u,k],{k,3}];
        u+dt(6 u(uplus[1]-u)/dx -
          (uplus[3]-3 uplus[2]+3 uplus[1]-u)/dx^3))
```

`plot2[i_:3]` lets us plot the result after i integration steps, using $dx = 0.05$ and $dt = 0.02$.

```
plot2[i_:3]:= (dx=0.05; dt=0.02; upast=ustart; time=0;
 Do[upres=step[upast];
    upast=upres;Print["Time ",time=time+dt],{i}] ;
    xulist = Table[{(j-max/2)dx,upres[[j]]},{j,0,max}];
    ListPlot[xulist, PlotJoined->True, PlotRange->All])
```

Figure 4.9 shows the result. After just three time steps, $u(x, t)$ exhibits oscillations, which are caused by numerical inaccuracies and "explode" after just two more steps. This, of course, is unphysical and can be blamed on the bad algorithm.

What went wrong? We have committed several errors, which we can avoid by the following modifications:

1. By a better discretization of the derivatives, the error in the approximation of $\partial/\partial t$ and $\partial^n/\partial x^n$ can be reduced.
2. The use of averages stabilizes the algorithm.
3. The step size dx chosen must always be large in comparison to dt.

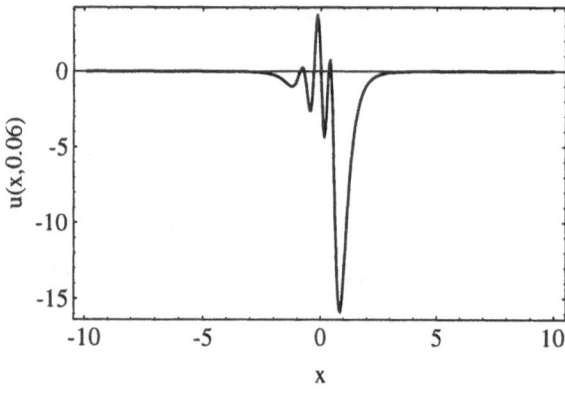

Fig. 4.9. Integration using **step[u]** and step sizes $dx = 0.05$, $dt = 0.02$. After just three integration steps, one obtains an unphysical result. The reader should compare this to Fig. 4.8

Point 1 can be illustrated by the following argument: Consider the Taylor expansion of $u(x,0)$ about $x = 0$, using the notation $u_0' = (\partial/\partial x)u(x,0)|_{x=0}$ etc.,

$$u_{\pm 1} = u_0 \pm dx u_0' + \frac{(dx)^2}{2} u_0'' \pm \frac{(dx)^3}{6} u_0''' + \cdots .$$

Then,

$$u_0' = \frac{u_1 - u_0}{dx} - \frac{dx}{2} u_0'' + \cdots ,$$

$$u_0' = \frac{u_1 - u_{-1}}{2 dx} - \frac{(dx)^2}{6} u_0''' + \cdots . \tag{4.45}$$

One sees that in the second case, u_0' can be calculated in such a way that $(dx/2)u_0''$ cancels exactly. This reduces the error by one order in dx. In a similar way, one can reduce the error of the higher derivatives by a suitable choice of the coefficients. One obtains

$$u_0'' = \frac{u_1 - 2u_0 + u_{-1}}{dx^2} + \mathcal{O}\left(dx^2\right) ,$$

$$u_0''' = \frac{u_2 - 2u_1 + 2u_{-1} - u_{-2}}{2 dx^3} + \mathcal{O}\left(dx^2\right) . \tag{4.46}$$

Thus, with respect to the quality of the approximation, the algorithm used in (4.44) gave away one order in dx and dt for all derivatives.

To illustrate points 2 and 3, we want to use a simple differential equation. In particular, we choose

$$\frac{\partial u}{\partial t} = -v \frac{\partial u}{\partial x}$$

with a constant velocity v. We discretize this equation in the form

$$\frac{u_{n+1}^j - u_n^j}{dt} = -v \frac{u_n^{j+1} - u_n^{j-1}}{2 dx} , \tag{4.47}$$

where we have already substituted the improved version (4.45) for $\partial/\partial x$. Solving for the components $(u_{n+1}^1, u_{n+1}^2, \ldots)$, which we combine into $\boldsymbol{u}_{n+1}$, we can write (4.47) as a matrix equation

$$\boldsymbol{u}_{n+1}^{\mathsf{T}} = \mathsf{M} \boldsymbol{u}_n^{\mathsf{T}} \tag{4.48}$$

with a tridiagonal matrix M. Obviously, the solution of (4.48) is $\boldsymbol{u}_n^{\mathsf{T}} = \mathsf{M}^n \boldsymbol{u}_0^{\mathsf{T}}$, from which we can see that the magnitude of the eigenvalues of M is critical for the stability of the algorithm. For the eigenmodes w of matrices of this type we use a Fourier ansatz of the form $w^j = \exp(ikj\,dx)$ with a wave vector k whose possible values are determined by the boundary conditions. If we choose an eigenmode $w = w_0$ as our initial state, and designate the corresponding eigenvalue as a, then applying the matrix M n times yields

$$w_n^j = a^n \exp\left(ikj\,dx\right) . \tag{4.49}$$

If we insert this into (4.47), we obtain an equation from which we can determine the amplitude a:

$$\frac{a^{n+1} - a^n}{dt} = -va^n \frac{e^{ikdx} - e^{-ikdx}}{2dx} \ . \tag{4.50}$$

The solution is

$$a = 1 - i\frac{vdt}{dx} \sin(kdx) \ . \tag{4.51}$$

Thus, a is complex, with a magnitude which is always greater than one, except for $k = 0$. This means that any initial mode which does not happen to be constant grows exponentially with time. The algorithm is unusable!

A seemingly insignificant change can stabilize the algorithm, however. In (4.47) we replace u_n^j by the average,

$$u_n^j \rightarrow \frac{1}{2} \left(u_n^{j+1} + u_n^{j-1} \right) \ . \tag{4.52}$$

This gives us

$$u_{n+1}^j = \frac{1}{2} \left(u_n^{j+1} + u_n^{j-1} \right) - \frac{vdt}{2dx} \left(u_n^{j+1} - u_n^{j-1} \right) \ , \tag{4.53}$$

and the amplitude equation yields

$$a = \cos(kdx) - i\frac{vdt}{dx} \sin(kdx) \ . \tag{4.54}$$

Now we have $|a| \leq 1$ for

$$|v|\,dt \leq dx \ . \tag{4.55}$$

This inequality is called the Courant condition. It states that perturbations only remain stable if the time step dt is chosen so that it is smaller than the propagation time $dx/|v|$. Or, from the opposite perspective, the spatial discretization chosen must not be too fine.

According to (4.45) it is better to discretize the time derivative symmetrically as well:

$$u_{n+1}^j - u_{n-1}^j = -\frac{vdt}{dx} \left(u_n^{j+1} - u_n^{j-1} \right) \ . \tag{4.56}$$

For the amplitude, this yields the equation

$$\left(a - \frac{1}{a} \right) = -2i\frac{vdt}{dx} \sin kdx \ . \tag{4.57}$$

The solution,

$$a = -i\frac{vdt}{dx} \sin kdx \pm \sqrt{1 - \left(\frac{vdt}{dx} \sin kdx \right)^2} \ , \tag{4.58}$$

shows that $|a| = 1$ if the Courant condition (4.55) holds. Here, too, dt has to be chosen to be sufficiently small compared to dx.

Equation (4.56) now contains three time steps, $n + 1$, n, and $n - 1$. To calculate the future, we have to know present and past. In the first step, we only know the present, so the future has to be calculated in one step, e.g., using (4.53). All other steps can then be done according to (4.56).

Now we want to use these insights for the numerical solution of the KdV equation. We substitute

$$\frac{\partial u}{\partial t} \rightarrow \frac{u_{n+1}^j - u_{n-1}^j}{2dt} \ ,$$

$$\frac{\partial u}{\partial x} \rightarrow \frac{u_n^{j+1} - u_n^{j-1}}{2dx} \ ,$$

$$u \rightarrow \frac{1}{3} \left(u_n^{j+1} + u_n^j + u_n^{j-1} \right) \ ,$$

$$\frac{\partial^3 u}{\partial x^3} \rightarrow \frac{1}{2(dx)^3} \left(u_n^{j+2} - 2u_n^{j+1} + 2u_n^{j-1} - u_n^{j-2} \right) \ .$$

We call $n - 1$ past, n present, and $n + 1$ future. Then a time step is given by

```
step2[u_,w_]:=(up1=RotateLeft[u]; up2=RotateLeft[up1];
               um1=RotateRight[u];um2=RotateRight[um1];
               w+dt(2(um1+u+up1)*(up1-um1)/dx -
                    (up2-2 up1+2 um1-um2)/dx^3 ) )
```

With step2, we generate ufut, the list for the future, if we insert the present (list upres) and the past (list upast) as arguments. By using plot3[i_:9], this is iterated and plotted after i time steps, a smooth curve through the x–u values having been drawn first via Interpolation[xulist].

```
plot3[i_:9]:=
    (Do[ufut=step2[upres,upast];
        upast=upres;upres=ufut; time = time+dt, {i}];
     Print["time ",time];
     xulist=Table[{(j-max/2)*dx,ufut[[j+1]]},{j,0,max}];
     uu=Interpolation[xulist];
     Plot[uu[x],{x,-10.,10.},PlotRange->All,
              Frame -> True, Axes -> None ] )
```

Figure 4.8 shows the result for the initial state $u(x,0) = -6 \operatorname{sech}^2(x)$. There is good agreement between the numerical solution and the exact one, which is shown by the dotted curve. The wave trough $u(x,0)$ turns into two solitons that move to the right at different velocities. This becomes particularly evident from the three-dimensional plot (Fig. 4.10) of $u(x,t)$ and the corresponding contour plot (Fig. 4.11). For negative times, the fast soliton with the large amplitude approaches the slow one. At $t = 0$, the two combine to form a single wave trough, and after the encounter both resume their original shape. While being overtaken, the small soliton experiences a delay, whereas the large one is accelerated. This can be seen particularly clearly in the contour plot of Fig. 4.11.

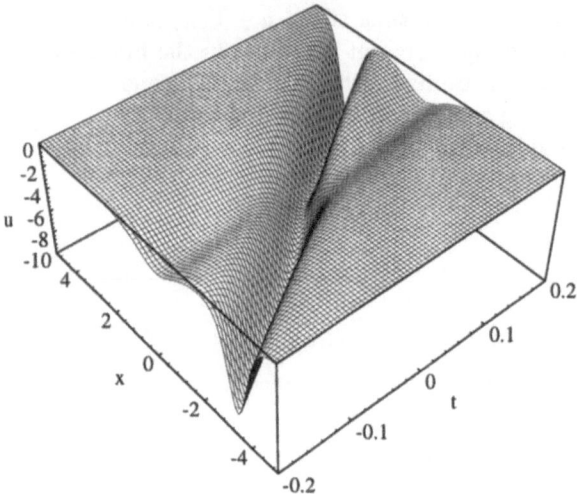

Fig. 4.10. The two solitons from Fig. 4.8, shown in a three-dimensional plot for different times. A large, fast soliton passes a small, broad one

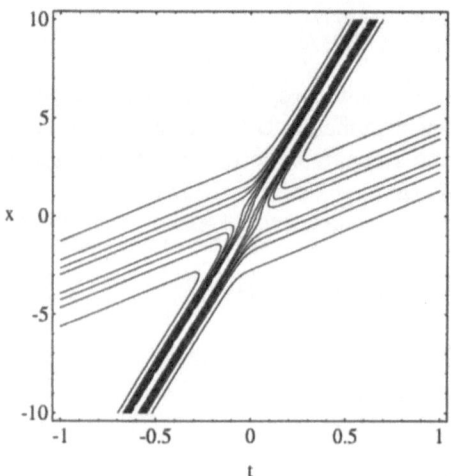

Fig. 4.11. The same data as in Fig. 4.10, but as a contour plot and with different ranges in space and time. While being passed, the small, broad soliton is delayed, whereas the large one is accelerated in the process

The analytic solutions mentioned at the beginning of this section evolve from an initial state $-N(N+1)\operatorname{sech}^2 x$, which decays to N solitons. Thus, the amplitudes -2 and -6 lead to the one- and two-soliton solution, respectively. But what happens to initial states whose amplitude falls in between? Figure 4.12 shows the numerical solution for the initial state $u(x,0) = -4\operatorname{sech}^2 x$. In addition to the soliton that moves to the right, additional waves disperse to the left.

It is self-evident that the quality of the numerical approximation can be improved by reducing the step size dx. The effort, however, increases signif-

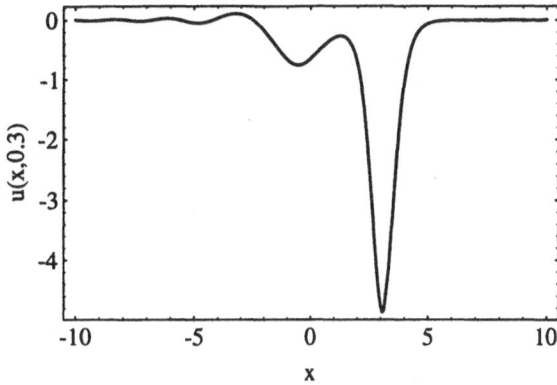

Fig. 4.12. Results similar to those in Fig 4.8, but for an initial state $u(x,0) = -4\,\mathrm{sech}^2 x$. In addition to the soliton moving to the right, waves are radiated

icantly in the process. Obviously, the length of the lists $\{u_n^j\}_{j=0}^{\max}$ increases as $1/dx$. The following exercise is intended to demonstrate that, at the same time, the maximum dt has to be reduced in proportion to $(dx)^3$, for the sake of stability.

Exercises

The stability of the algorithm programmed in step2[u_,w_] is limited significantly by the discretized third derivative.

1. Remove the nonlinear term from the full KdV equation, i.e., investigate the equation

$$\frac{\partial u}{\partial t} = -\frac{\partial^3 u}{\partial x^3} . \tag{4.59}$$

The discretization used here yields

$$u_{n+1}^j = u_{n-1}^j - \frac{dt\left(u_n^{j+2} - 2u_n^{j+1} + 2u_n^{j-1} - u_n^{j-2}\right)}{(dx)^3} . \tag{4.60}$$

Investigate the stability of this algorithm by using the ansatz (4.49) in (4.60). Prove that, analogously to (4.57), this leads to the amplitude equation

$$a = \frac{1}{a} - \frac{2idt\left[\sin\left(2kdx\right) - 2\sin\left(kdx\right)\right]}{(dx)^3} .$$

Numerically determine the maximum μ of $|\sin(2kdx) - 2\sin(kdx)|$ ($\mu \simeq 2.6$), and show that, in combination with the requirement $|a| \le 1$, this yields the stability condition

$$dt \le \frac{1}{\mu}(dx)^3 . \tag{4.61}$$

2. Use the algorithm step2[u,w] for the integration, and choose the initial condition $u(x,0) = -6\,\text{sech}^2 x$. Vary d$x$ between 0.1 and 0.2 and determine the stability limit with respect to dt, for a given dx, i.e., approximately determine the maximum dt allowed as a function of dx. Confirm (4.61) by displaying the result in a log–log plot.

Literature

Baumann G. (1996) Mathematica in Theoretical Physics: Selected Examples from Classical Mechanics to Fractals. TELOS, Santa Clara, CA

Crandall R.E. (1991) Mathematica for the Sciences. Addison-Wesley, Redwood City, CA

4.5 Time-dependent Schrödinger Equation

If a particle moves in a box without friction or any other force, its motion is changed only by reflection off the walls. In one dimension, therefore, it moves back and forth regularly. This classical picture, which is based on the idea of a pointlike mass with precisely defined position and momentum, no longer holds true if the box is microscopically small. Instead, we have to describe the particle by a wave function. In the quantum mechanics course one learns that the stationary states in this case are standing waves. But what happens to an initially localized wave packet that moves towards the walls of the box?

This problem has both analytic and numerical aspects. We will use the expansion in terms of eigenfunctions in order to make statements about symmetries, recurrence times, and other characteristic length and time scales. On the other hand, we want to use the time-dependent Schrödinger equation to demonstrate how to solve partial differential equations numerically by using an implicit method. This involves inverting a tridiagonal matrix, for which there is a fast numerical method. Additionally, care has to be taken when discretizing the time evolution of the quantum-mechanical state that the normalization of the wave function does not change with time.

We will see that the seemingly simple and well-understood textbook example exhibits a surprisingly complex behavior. The wave packet disperses, and wild interference patterns arise from which smooth wave packets suddenly reemerge. Finally, the entire process repeats periodically with time. The cover of this book shows what complexity and beauty can arise from elementary quantum mechanics.

Physics

Let $\psi(x,t)$ be the complex-valued wave function of a particle with mass m which moves in one dimension in a potential $\tilde{V}(x)$. Then $|\psi(x,t)|^2 dx$ is the probability of finding the particle in the interval $[x, x + dx]$ at time t. The time dependence of $\psi(x, t)$ is described by the Schrödinger equation in its coordinate representation:

$$i\hbar \frac{\partial \psi}{\partial t} = -\frac{\hbar^2}{2m} \frac{\partial^2 \psi}{\partial x^2} + \tilde{V}(x)\psi . \tag{4.62}$$

In order to put this equation into a dimensionless form, we normalize the time by t_0 and the position by x_0:

$$i\frac{\hbar}{t_0} \frac{\partial \psi}{\partial (t/t_0)} = -\frac{\hbar^2}{2mx_0^2} \frac{\partial^2 \psi}{\partial (x/x_0)^2} + \tilde{V}(x)\psi . \tag{4.63}$$

Now we choose t_0 and x_0 such that the following equation holds:

$$\hbar t_0 = 2mx_0^2 . \tag{4.64}$$

Now if we set $\tilde{V}(x)t_0/\hbar = V(x/x_0)$ and express position and time in units of x_0 and t_0, we obtain the dimensionless equation

$$i\frac{\partial \psi}{\partial t} = H\psi = \left[-\frac{\partial^2}{\partial x^2} + V(x) \right] \psi . \tag{4.65}$$

H is the normalized Hamiltonian of the particle. There are two ways to numerically solve this equation. First, we can calculate the eigenstates and eigenvalues of the stationary equation $H\psi = E\psi$, expand $\psi(x, 0)$ in terms of these states, and then specify a series representation for $\psi(x, t)$. Second, we can directly integrate the time-dependent equation as shown in the previous section. The second method can even be used for problems for which the first one fails, which is why we want to describe that method in the algorithm section. If, on the other hand, the eigenstates are known, one can use the analytic ansatz to directly derive at least some properties of the wave function $\psi(x, t)$.

We want to study a particle in a box with infinitely high walls. As symmetry considerations will play a significant role in what follows, we position the coordinate system in such a way that this symmetry can be expressed easily. With these considerations, the potential $V(x)$ takes the form

$$V(x) = \begin{cases} 0 & \text{for} \quad -\frac{1}{2} \leq x \leq \frac{1}{2} , \\ \infty & \text{otherwise} . \end{cases} \tag{4.66}$$

Here, the coordinate x is expressed in units of the box's width a, and the energy E in units of $\hbar^2/2ma^2$. The energies E_n of the stationary states $\psi_n(x)$ are known from the quantum mechanics course:

$$E_n = n^2 \pi^2 , \quad \text{where} \quad n = 1, 2, 3, \dots ,$$
$$\psi_n(x) = \begin{cases} \sqrt{2}\cos(n\pi x) & \text{for } n \text{ odd} \\ \sqrt{2}\sin(n\pi x) & \text{for } n \text{ even} \end{cases} , \quad -\frac{1}{2} \leq x \leq \frac{1}{2} . \tag{4.67}$$

These states are standing waves whose magnitude does not change as a function of time. The particle's mean position in these states, $\langle x \rangle = 0$, does not change with time either.

We now want to study a wave packet $\psi(x,t)$ that moves back and forth within the box. The function $\psi(x,0)$ can be expanded in terms of eigenstates, so for $\psi(x,t)$ we obtain

$$\psi(x,t) = \sum_{m=1}^{\infty} \hat{c}_{2m-1} \exp\left[i\pi^2 (2m-1)^2 t\right] \cos\left[(2m-1)\pi x\right]$$

$$+ \sum_{m=1}^{\infty} \hat{c}_{2m} \exp\left[-i\pi^2 (2m)^2 t\right] \sin(2m\pi x) \tag{4.68}$$

$$= \psi_e(x,t) + \psi_o(x,t) \tag{4.69}$$

with expansion coefficients $\{\hat{c}_n\}$. Owing to the symmetry of the basis functions, this expansion yields at the same time the decomposition of $\psi(x,t)$ into its even (ψ_e) and odd (ψ_o) parts.

Thus the general solution of the Schrödinger equation is an infinite superposition of standing waves with the phase factor of each wave changing at a different rate. This gives rise to complex interference patterns that can change very quickly with time (see the results section below). The states in the box have one distinctive feature, however, that is related to the structure of the energies E_n. Each frequency $\omega_n = n^2\pi^2$ is an integer multiple of the fundamental frequency $\omega_1 = \pi^2$. From this it follows that $\psi(x,t)$ is a periodic function of time, with period $T = 2\pi/\omega_1 = 2/\pi$. After the time $t = T$ has elapsed, the wild interference patterns vanish and the wave packet $\psi(x,t)$ entirely returns to its initial state $\psi(x,0)$.

We will see in the results section that simple patterns emerge even at certain earlier times. For example, after a time $t = T/2$ the initial distribution reemerges, reflected about the center of the box, moving exactly in the opposite direction. We want to present a brief analytic proof of this. At the time $t = T/2 = 1/\pi$, all phase factors in $\psi_e(x,t)$ give the factor -1, since $\exp(-i\pi(2m-1)^2) = -1$, whereas the exponentials in $\psi_o(x,t)$ all have the value 1 for $t = 1/\pi$. Consequently, we obtain

$$\psi\left(x, \tfrac{T}{2}\right) = -\psi_e(x,0) + \psi_o(x,0)$$

$$= -\psi_e(-x,0) - \psi_o(-x,0) = -\psi(-x,0) . \tag{4.70}$$

Thus after the time $T/2$ the initial probability distribution $|\psi(x,0)|^2$ is back, but it is reflected about the center of the box, $x = 0$. That the wave moves in the opposite direction can best be seen from the current density,

$$j(x,t) = i\left[\psi(x,t)\frac{\partial \psi^*}{\partial x}(x,t) - \psi^*(x,t)\frac{\partial \psi}{\partial x}(x,t)\right] . \tag{4.71}$$

If we insert the result from (4.70) here, we immediately get

$$j\left(x, \tfrac{T}{2}\right) = -j(-x,0) . \tag{4.72}$$

We have shown that at the times $T/2$ and T the original form reemerges from the complex interference patterns. The results section shows, however, that simple waves arise even earlier. For example, at $T/4$ we get two wave packets which move towards one another and interfere with each other later. This is a consequence of the initial condition we have chosen,

$$\psi(x,0) = e^{ikx} \exp\left[-\frac{1}{2\sigma^2}\left(x+\frac{1}{4}\right)^2\right] . \tag{4.73}$$

The function ψ is a Gaussian of width σ, which is concentrated near $x = -1/4$, multiplied by $\exp(ikx)$. Owing to this factor, the particle has an initial velocity

$$\langle v \rangle_0 = \frac{\langle \psi_{t=0} | (p/m) | \psi_{t=0} \rangle}{\|\psi\|^2} = \frac{\hbar k}{m} = 2k . \tag{4.74}$$

The width σ has to be sufficiently small that the boundary condition $\psi(\pm 1/2, 0) = 0$ can be regarded as fulfilled to sufficient precision. For the numerical simulation we use $\sigma = 0.05$, i.e., $|\psi(\pm 1/2, 0)| \leq \exp(-12.5) \lesssim 4 \cdot 10^{-6}$ as compared to $|\psi(-1/4, 0)| = 1$.

We insert $t = T/4 = 1/(2\pi)$ into (4.68). Now the phase factors yield $\exp[-i\pi^2(2m-1)^2/(2\pi)] = -i$ and $\exp[-i\pi^2(2m)^2/(2\pi)] = 1$ respectively, and we obtain

$$\psi\left(x, \frac{T}{4}\right) = -i\psi_e(x,0) + \psi_o(x,0) . \tag{4.75}$$

Consequently, the norm of the wave function for $t = T/4$ is

$$\left|\psi\left(x, \frac{T}{4}\right)\right|^2 = |\psi_e(x,0)|^2 + |\psi_o(x,0)|^2$$
$$+ i\left[\psi_e^*(x,0)\psi_o(x,0) - \psi_e(x,0)\psi_o^*(x,0)\right] . \tag{4.76}$$

The last term of this equation, which can be written as $2\,\mathrm{Im}\,(\psi_e(x,0)\psi_o^*(x,0))$, is negligibly small. A simple calculation using (4.73) yields

$$2\,\mathrm{Im}\,[\psi_e(x,0)\psi_o^*(x,0)] = -2\exp\left[-\frac{1}{2\sigma^2}\left(\frac{1}{8} + 2x^2\right)\right]\cos kx \sin kx , \tag{4.77}$$

i.e., a term which for $\sigma = 0.05$ is of the order $\exp(-25)$. The remainder can be expressed by $\psi(\pm x, 0)$ in the following way:

$$|\psi_e(x,0)|^2 + |\psi_o(x,0)|^2$$
$$= \frac{1}{2}\left[|\psi_e(x,0) + \psi_o(x,0)|^2 + |\psi_e(x,0) - \psi_o(x,0)|^2\right]$$
$$= \frac{1}{2}\left[|\psi_e(x,0) + \psi_o(x,0)|^2 + |\psi_e(-x,0) + \psi_o(-x,0)|^2\right]$$
$$= \frac{1}{2}\left[|\psi(x,0)|^2 + |\psi(-x,0)|^2\right] .$$

In summary we obtain

$$\left|\psi\left(x, \frac{T}{4}\right)\right|^2 \cong \frac{1}{2}\left[|\psi(x,0)|^2 + |\psi(-x,0)|^2\right] , \tag{4.78}$$

i.e., just a symmetrized version of the initial distribution. The corresponding calculation for the current density yields

$$j\left(x, \tfrac{T}{4}\right) \cong \frac{1}{2} j\left(x, 0\right) - \frac{1}{2} j\left(-x, 0\right) , \tag{4.79}$$

which shows that the two wave packets (4.78) move towards each other.

At even shorter times, waves with a simple structure arise as well. If we set $t = Tp/q$ with integers p and q which are relatively prime, the phase $\exp(-2\pi i n^2 p/q)$ can take at most q different values. Therefore, according to (4.68), $\psi(x, t)$ is a superposition of q waves, each of which is an infinite partial sum out of that same expansion for $\psi(x, 0)$. If we assume that each partial wave is a smooth function, then their superposition is smooth as well and $\psi(x, t)$ exhibits a simple structure at those times. This is precisely what we will observe when we do the numerical integration.

With this we conclude the analytic considerations and turn our attention to the description of the numerical integration of the time-dependent Schrödinger equation.

Algorithm

First, $\psi(x, t)$ is discretized:

$$\psi_n^j = \psi\left(j \mathrm{d}x, n \mathrm{d}t\right) . \tag{4.80}$$

Here, j and n are integers, and $\mathrm{d}x$ and $\mathrm{d}t$ are the step sizes of the space and time coordinates. As before, we write the second derivative as

$$\frac{\partial^2 \psi}{\partial x^2} \to \frac{1}{(\mathrm{d}x)^2} \left(\psi_n^{j+1} - 2\psi_n^j + \psi_n^{j-1}\right) \tag{4.81}$$

with an error of the order $\mathcal{O}(\mathrm{d}x^2)$. Now we can discretize the time derivative, analogously to the solitons from the previous section. That algorithm, however, does not conserve the overall probability $\int |\psi(x, t)|^2 \mathrm{d}x = 1$, which is not allowed to change with time. It is indeed possible to find a discrete approximation of the partial differential equation that conserves the overall probability. To this end, the discretized time evolution operator must be chosen to be unitary.

The solution of the Schrödinger equation (4.65) can also be written in the form

$$\psi\left(x, t + \mathrm{d}t\right) = \mathrm{e}^{-\mathrm{i}H\mathrm{d}t} \psi\left(x, t\right) . \tag{4.82}$$

The approximation

$$\mathrm{e}^{-\mathrm{i}H\mathrm{d}t} \simeq (1 - \mathrm{i}H\mathrm{d}t) + \mathcal{O}\left(\mathrm{d}t^2\right) , \tag{4.83}$$

which corresponds to the simple Euler step, is no longer unitary. On the other hand, with

$$e^{-iHdt} = \left(e^{iHdt/2}\right)^{-1} e^{-iHdt/2}$$

$$\simeq \left(1 + \frac{i}{2}Hdt\right)^{-1} \left(1 - \frac{i}{2}Hdt\right) + \mathcal{O}\left(dt^3\right) \tag{4.84}$$

one obtains a unitary operator for the discrete time evolution:

$$\psi_{n+1}^j = \left(1 + \frac{i}{2}Hdt\right)^{-1} \left(1 - \frac{i}{2}Hdt\right) \psi_n^j . \tag{4.85}$$

Now the overall probability $\sum_j |\psi_n^j|^2$ remains constant as a function of time n. The inverse operator can be moved to the left side of the equation just as well,

$$\left(1 + \frac{i}{2}Hdt\right) \psi_{n+1}^j = \left(1 - \frac{i}{2}Hdt\right) \psi_n^j . \tag{4.86}$$

In this form, the difference equation becomes an implicit equation; this means that the wave function in the next time step is determined by a system of equations that has to be solved first. The corresponding algorithm takes more processing time, but it leads to a more stable numerical procedure in almost all cases. If we now insert the particular form of the Hamiltonian, we obtain

$$(H\psi)_n^j = -\frac{1}{(dx)^2} \left(\psi_n^{j+1} - 2\psi_n^j + \psi_n^{j-1}\right) + V^j \psi_n^j . \tag{4.87}$$

With this, (4.86) yields

$$\psi_{n+1}^{j+1} + \left(i\frac{2(dx)^2}{dt} - (dx)^2 V^j - 2\right) \psi_{n+1}^j + \psi_{n+1}^{j-1} =$$

$$- \psi_n^{j+1} + \left(i\frac{2(dx)^2}{dt} + (dx)^2 V^j + 2\right) \psi_n^j - \psi_n^{j-1} = \Omega_n^j . \tag{4.88}$$

We have abbreviated the second line, which only contains terms referring to the time step n, by Ω_n^j. This equation has the form

$$\mathsf{T}\psi_{n+1} = \Omega_n , \tag{4.89}$$

with a tridiagonal matrix T. Therefore we have to solve a system of linear equations. In the case of tridiagonal matrices, there is a special method for this. We use the ansatz

$$\psi_{n+1}^{j+1} = a^j \psi_{n+1}^j + b_n^j . \tag{4.90}$$

If we insert this in (4.88), we find

$$\psi_{n+1}^j = \left[2 + (dx)^2 V^j - i\frac{2(dx)^2}{dt} - a^j\right]^{-1} \left[\psi_{n+1}^{j-1} + b_n^j - \Omega_n^j\right] . \tag{4.91}$$

Comparison with (4.90) yields equations for a^{j-1} and b_n^{j-1}, which we can solve for a^j and b_n^j. The result is

$$a^j = 2 + (dx)^2 V^j - i \frac{2 (dx)^2}{dt} - \frac{1}{a^{j-1}} \,,$$

$$b_n^j = \Omega_n^j + \frac{b_n^{j-1}}{a^{j-1}} \,. \qquad (4.92)$$

As required by the ansatz (4.90), a^j is independent of n, whereas b_n^j is calculated from the wave function ψ_n^j. To solve these equations, we need the initial state ψ_0^j and boundary conditions. Since we want to confine the particle in a box with infinitely high walls, ψ has to vanish at the boundary,

$$\psi_n^0 = 0 \quad \text{and} \quad \psi_n^J = 0 \,, \qquad (4.93)$$

where we have redefined the boundary as $x = 0$ and $x = 1$, as this is easier to program. The number of grid points, J, is thus fixed at $J = 1 + 1/dx$. At $j = 1$, (4.88) becomes

$$\psi_{n+1}^2 + \left(i \frac{2 (dx)^2}{dt} - (dx)^2 V^1 - 2 \right) \psi_{n+1}^1 = \Omega_n^1 \,. \qquad (4.94)$$

A comparison with (4.90) gives

$$a^1 = 2 + (dx)^2 V^1 - i \frac{2 (dx)^2}{dt} \,, \quad b_n^1 = \Omega_n^1 \,. \qquad (4.95)$$

From these initial values, all a^j can be calculated using (4.92), as can all b_n^j after each time step. The wave function is then calculated from (4.90) by inverse iteration:

$$\psi_{n+1}^j = \frac{\psi_{n+1}^{j+1} - b_n^j}{a^j} \,. \qquad (4.96)$$

Since the boundary value on the right, $\psi_{n+1}^J = 0$, is fixed, this allows the calculation of the entire vector ψ_{n+1}^j at the time step $n + 1$.

All results are now combined into one algorithm:

1. Choose an initial state ψ_0^j and use it to calculate Ω_0^j according to (4.88).
2. Calculate the vector a^j from the initial value given in (4.95) and the recursion formula (4.92).
3. For all positions j, calculate the variable b_n^j, using the initial value from (4.95) and the recursion formula (4.92).
4. Use (4.96) to calculate the values of the wave function ψ_{n+1}^j for the next time step, using the initial value $\psi_{n+1}^J = 0$. This also yields Ω_{n+1}^j according to (4.88).
5. Repeat steps 3 and 4 for $n = 0, 1, 2, \ldots$.

As mentioned before, we define the initial state as a Gaussian wave packet that moves with a mean momentum k:

$$\psi(x, 0) = e^{ikx} \exp \left[-\frac{(x - x_0)^2}{2\sigma^2} \right] \,. \qquad (4.97)$$

The equations above are easily formulated in any programming language. In doing so, it has to be noted, however, that the quantities $\{\psi_n^j\}$, $\{a^j\}$, and $\{b_n^j\}$ are complex numbers. Whereas *Mathematica*, as well as other languages like Fortran, can deal with those numbers directly, in C one has either to include the appropriate routines, e.g., from *Numerical Recipes*, or to define a structure with a real and imaginary part oneself:

```
typedef struct{double real, imag; } complex;
```

In this case, one also has to write multiplication and division routines oneself. For example, the well-known multiplication and division formulae for complex numbers $a = \operatorname{Re} a + i \operatorname{Im} a$ and $b = \operatorname{Re} b + i \operatorname{Im} b$,

$$a \cdot b = \operatorname{Re} a \operatorname{Re} b - \operatorname{Im} a \operatorname{Im} b + i \left(\operatorname{Im} a \operatorname{Re} b + \operatorname{Re} a \operatorname{Im} b \right) , \tag{4.98}$$

$$\frac{a}{b} = \frac{\operatorname{Re} a \operatorname{Re} b + \operatorname{Im} a \operatorname{Im} b + i \left(\operatorname{Im} a \operatorname{Re} b - \operatorname{Re} a \operatorname{Im} b \right)}{\left(\operatorname{Re} b \right)^2 + \left(\operatorname{Im} b \right)^2} \tag{4.99}$$

can be found in the function `calculate_b` which performs the iteration of (4.92)

```
void calculate_b(complex *b, complex *omega, complex *a)
{
int j ; double a2;

b[1].real=omega[1].real;
b[1].imag=omega[1].imag;
for (j=2;j<J;j++)
 {a2=a[j-1].real*a[j-1].real+a[j-1].imag*a[j-1].imag;
  b[j].real=omega[j].real+
  (b[j-1].real*a[j-1].real+b[j-1].imag*a[j-1].imag)/a2;
  b[j].imag=omega[j].imag+
  (b[j-1].imag*a[j-1].real-b[j-1].real*a[j-1].imag)/a2;
 }
}
```

All other steps of the algorithm above can be programmed just as easily. We refer the reader to the source code on the enclosed CD-ROM.

To numerically integrate the Schrödinger equation we have chosen the interval $[0, 1]$ for the box, because, among other things, this makes it easier to convert the x-coordinate to the positive screen coordinates. The eigenfunction expansion that corresponds to (4.68) is then given by:

$$\psi (x,t) = \sum_{n=1}^{\infty} c_n \exp \left(-in^2\pi^2t \right) \sin \left(n\pi x \right) , \tag{4.100}$$

$$c_n = 2 \int_0^1 \sin \left(n\pi x \right) \psi (x,0) \, dx . \tag{4.101}$$

The initial state in our simulation is the Gaussian wave packet (4.97) with a width $\sigma = 0.05$ and the mean momentum $k = 40$ which is centered at $x_0 = 0.25$. As we have already seen, the wave has to return to its initial state after the time $T = 2/\pi \simeq 0.6366$. This provides a test of the accuracy with which we have integrated the Schrödinger equation.

Of course, T is also the largest time scale that is relevant for this problem. In addition, though, there are three other characteristic time scales that play a role in this case. First, there is the time it takes the particle to traverse the box once, given its momentum. For $k = 40$, we obtain $t_1 = 1/80 = 0.0125$ from (4.74). Next, we have to take into account the dispersion of the wave packet, which is a typical effect of quantum mechanics. A free Gaussian distribution with an initial width σ will have the width $\sigma\sqrt{1 + t^2\hbar^2/(4m^2\sigma^4)}$ at time t. By asking when this width will agree with the size of the box, we obtain the time scale $t_2 \cong 1/20 = 0.05$.

Finally, the fourth time scale, which also gives us a handle on selecting the step sizes dx and dt, is determined by the highest relevant phase velocity, for whose determination we make use of (4.100) and (4.101). If we express the sine in these equations by an exponential function, the individual terms in the sum (4.100) take the form

$$c_n \exp\left[-in\pi\left(n\pi t \pm x\right)\right] , \tag{4.102}$$

so the phase velocity v_n^{ph} for this partial wave has the value

$$v_n^{\mathrm{ph}} = \pm n\pi . \tag{4.103}$$

To estimate up to which n the coefficients c_n give significant contributions, we note that the Fourier transform of a Gaussian is a Gaussian again, but with the reciprocal width. The factor $\exp(ikx)$ turns into a shift by the distance k in Fourier space. From this, we can derive the n-dependence of c_n

$$|c_n| \propto \exp\left[-\frac{1}{2}\sigma^2\left(n\pi - k\right)^2\right] . \tag{4.104}$$

On the screen, we can only display those amplitudes which amount to at least $1/1000$ of the maximum amplitude. This leads us to the estimate

$$\frac{1}{2}\sigma^2\left(n\pi - k\right)^2 \leq \ln\left(1000\right) \quad \Rightarrow \quad n_{\mathrm{max}} \simeq 35 , \quad v_{\mathrm{max}}^{\mathrm{ph}} \simeq 100 . \tag{4.105}$$

With the requirement that the shortest arc among those sines be covered by about 20 to 30 steps, this consideration leads to

$$dx \lesssim 10^{-3} \quad \text{and} \quad dt \lesssim 10^{-5} , \tag{4.106}$$

since dt should not be larger than $dx/v_{\mathrm{max}}^{\mathrm{ph}}$ in any case.

If one also wants to take into account the global aspect that the integration should be correct for at least the duration of a period, the step size dt can be determined more accurately by the following consideration. While the algorithm we use guarantees that the norm of the wave function is conserved, it still represents an approximation. For while the exact time evolution of the eigenmode $\psi_n(x)$ after ℓ steps is given by the factor $\exp(-in^2\pi^2\ell dt)$, the numerical calculation according to (4.85) yields

$$\left(\frac{1-in^2\pi^2 \, dt/2}{1+in^2\pi^2 \, dt/2}\right)^{\ell} = e^{i\ell d\varphi} \ , \quad \text{where} \quad d\varphi = 2\arctan\left(n^2\pi^2\frac{dt}{2}\right) .\quad(4.107)$$

For the absolute value of the difference between these two expressions, the Taylor expansion of $d\varphi$ gives the result

$$\Delta \simeq \frac{1}{12}\ell\left(dtn^2\pi^2\right)^3 . \tag{4.108}$$

If we require that $\Delta < 1/2$ for $\ell = T/dt$ and $n = n_{\max}$, we obtain a step $dt \lesssim 2\cdot 10^{-6}$. It turns out that for $dx = 10^{-3}$ and $dt = 5\cdot 10^{-6}$ our algorithm produces reasonable results which do, however, exhibit deviations from the exact time evolution as the integration time increases.

Results

Initially, the wave does not yet notice the box. It moves to the right and disperses in the process. As soon as part of the wave is reflected off the wall, though, it interferes with the part that is still coming in, forming a wave pattern (Fig. 4.13, left). In the right-hand part of the figure one can see that after just a short time ($t = T/40$) an irregular interference pattern has formed that spans the entire box. This pattern is typical for almost all times.

Suddenly, though, regular shapes arise from the seemingly chaotic movements, as can be seen in Figs. 4.14 and 4.15 for the times $T/12$, $T/4$, $T/3$, and $T/2$. As shown before, two wave packets form at $t = T/4$, which move towards and then interfere with one another (Fig. 4.14, right), and at $t = T/2$ the original wave suddenly reemerges, reflected about the center of the box and moving to the left (Fig. 4.15, right). These simple patterns, however, quickly disperse and interfere again to generate wild movements.

In Fig. 4.16, the individual pictures for different times are combined into a mountain range rising above the x–t plane. The spatial coordinate is displayed horizontally from right to left and time from back to front. At the back, one can see how the initial state first moves to the left, dispersing in the

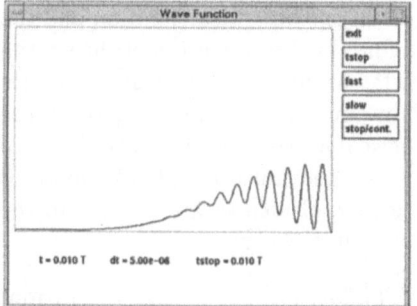 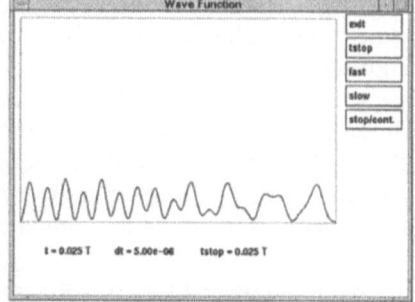

Fig. 4.13. The wave packet $|\psi(x,t)|^2$ shortly after the first impact on the wall ($t = T/100$, *left*) and a typical interference pattern $|\psi(x,t)|^2$ for $t = T/40$ (*right*)

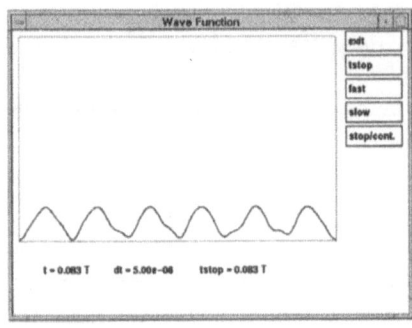 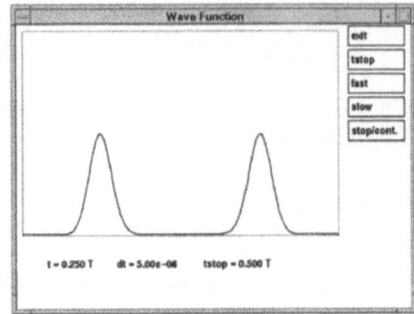

Fig. 4.14. The wave packet $|\psi(x,t)|^2$ at times $t = T/12$ *(left)* and $t = T/4$ *(right)*

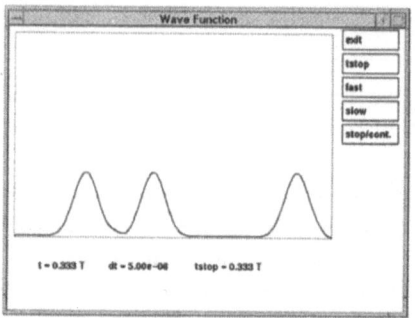 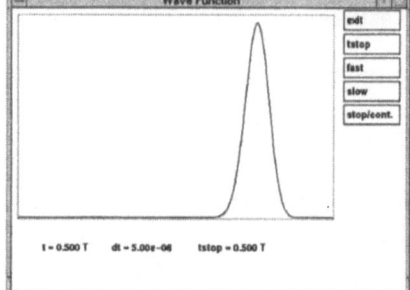

Fig. 4.15. The wave packet $|\psi(x,t)|^2$ at times $t = T/3$ *(left)* and $t = T/2$ *(right)*

process, and then is reflected off the wall. In the foreground, the probability distribution just before the time $T/6$ can be seen: a smooth wave with three peaks. In between, a fascinating hilly landscape of unexpected variety and regularity arises. In particular the valleys, which spread out in star shapes, did not attract our attention while studying the wave directly on the screen. So far, we have not been able to calculate these valleys analytically.

Finally, Fig. 4.17 shows both the initial distribution $|\psi(x,0)|^2$ (dotted) and the numerically integrated wave packet after a full period T. If we wanted to eliminate the obvious discrepancy between the numerical solution and the exact time evolution, we would have to refine the discretization of the spatial coordinate even more. This is because so far we have used the energy values $E_\nu = \nu^2\pi^2$ of the exact Hamiltonian H in our considerations regarding the duration of a period, whereas for the numerical integration the matrix version of H, (4.87) with $V^j = 0$, is relevant. The eigenvalues E_ν^{d} of this operator and the corresponding eigenvectors ϕ_ν^j, which, of course, are subject to the boundary conditions (4.93), are known as well, however:

$$\phi_\nu^j = \sin\left(\nu\pi j\mathrm{d}x\right) \ , \quad E_\nu^{\mathrm{d}} = \frac{4}{\left(\mathrm{d}x\right)^2}\sin^2\left(\frac{\nu\pi}{2}\mathrm{d}x\right) \ . \tag{4.109}$$

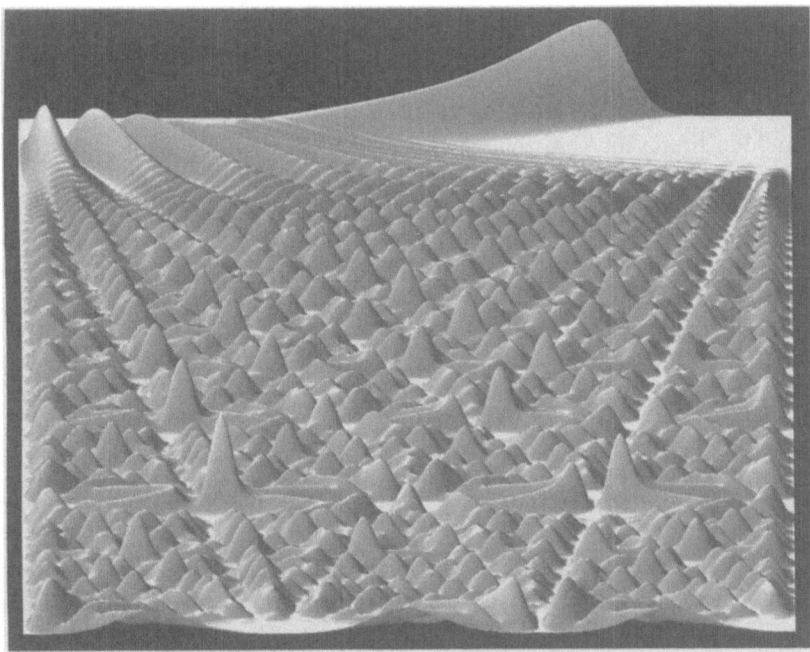

Fig. 4.16. Space–time diagram of the probability of presence of a quantum particle in a square-well potential

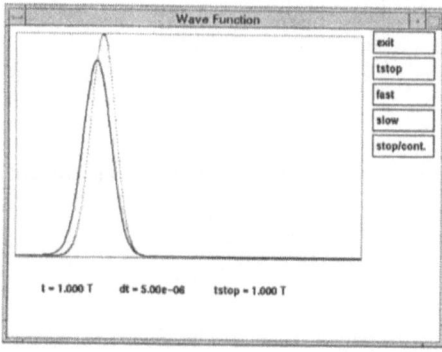

Fig. 4.17. The initial distribution $|\psi(x,0)|^2$ (*dotted*) and the numerically integrated wave packet $|\psi(x,T)|^2$ after the full period T (*solid curve*)

From the equation above and the expansion of E_ν^d for $dx \ll 1$,

$$E_\nu^d \simeq \nu^2 \pi^2 - \frac{1}{12} \nu^4 \pi^4 (dx)^2 \; , \tag{4.110}$$

we conclude first that the frequencies no longer are integer multiples of a fundamental frequency, and second that all of them are smaller than the E_ν previously considered. This explains both the broadening and the time delay of the wave packet in Fig. 4.17. By taking the modified energy values E_ν^d according to (4.109) into account in (4.107) one obtains $dx \lesssim 10^{-4}$ and $dt \lesssim 10^{-6}$ if one does not want this correction to be noticeable after one period.

Exercises

- Use the integration routine described in this section to calculate and display the mean values $\langle x \rangle$, $\langle p \rangle$, and their uncertainties Δx, Δp as a function of time.
- The program code above is easily modified to demonstrate the tunnel effect. Change the potential inside the box by putting a Gaussian barrier of the form

$$V_0 \exp \left[\frac{-(x - 1/2)^2}{2d^2} \right]$$

in its center. Try finding a numerical criterion for the tunneling time t_T. For example, the probability w_n of finding the particle on the right-hand side of the barrier at the time $t_n = ndt$ is given by

$$w_n = \frac{\sum_{j=J/2}^{J} |\psi_n^j|^2}{\sum_{j=1}^{J} |\psi_0^j|^2} \; .$$

Numerically determine the dependence of the tunneling time V_0 on the width d of the barrier.

Literature

Dahmen H.D., Brandt S. (1994) Quantum Mechanics on the Personal Computer. Springer, Berlin, Heidelberg, New York

Koonin S.E., Meredith D.C. (1990) Computational Physics, Fortran Version. Addison-Wesley, Reading, MA

5. Monte Carlo Simulations

Monte Carlo → casino → roulette → random numbers: This is the chain of associations which gave an important method of computer simulation its name. With the help of random numbers, one can use the computer to simulate, for example, the motion of an interacting many-body system in a heat reservoir. As in the real experiment the temperature and other parameters can be varied. The materials being modeled can be heated up or cooled down, and at sufficiently low temperatures one can observe how gases liquefy, how atoms in a magnetic material get aligned, or how metals lose their electric resistance.

Here we want to use simple examples to study how interesting physics phenomena can be described with the help of random numbers. Some of these models even have universal properties: The values of the critical exponents that describe the singularities at phase transitions are the same for many different models and are even measured in real materials. Therefore computer simulation is a particularly important tool for understanding the cooperative properties of interacting particles.

5.1 Random Numbers

A computer cannot generate random numbers. It works according to a well-defined program, i.e., according to rules which are applied to input data and generate output data. Therefore a computer acts like a deterministic function that leaves no room for chance. Still, there are algorithms that generate "pseudo-random" numbers. In many statistical tests, such a sequence of numbers leads to results similar to those we would get from random numbers that fulfill the mathematical definition of "randomness." We want to briefly introduce such algorithms here.

Algorithm and Results

In a computer, numbers are represented by a sequence of bits (0 or 1). If, for example, 32 bits per number are available, a maximum of 2^{32} different numbers can be represented with these bits. Consequently, a function f acting on these numbers,

$$r_n = f(r_{n-1}) \; , \qquad\qquad\qquad (5.1)$$

generates a sequence $r_0, r_1, r_2, \ldots$ which must repeat itself after at most 2^{32} steps. Therefore a computer can only generate periodic sequences of numbers. If the period attains its maximum, each number appears with the same frequency in a very long sequence, i.e., the numbers are uniformly distributed. If the sequence of numbers generated by the computer passes many tests for randomness, it is called a sequence of random numbers.

One function $f(r)$ that is frequently used in this context is a linear function followed by the modulo operation with three parameters a, c, and m:

$$r_n = (a\,r_{n-1} + c) \bmod m \; . \qquad\qquad (5.2)$$

If one calls the random number generators provided by the system, e.g., rand() in C or Random[] in *Mathematica*, these so-called linear congruential generators are almost always used. The result of the modulo function (= remainder when dividing by m) is limited to at most m different values; therefore, the maximum period is m. For some values of the parameters a and c, this maximum m is actually reached. Some generators with such parameters have fared relatively well in tests for randomness. Some favorable values for the parameters (m, a, c) are given in *Numerical Recipes*, e.g., $m = 6075$, $a = 106$, and $c = 1283$. We have used this generator to generate points on the surface of the three-dimensional unit sphere from triples of consecutive random numbers,

$$e_n = \frac{(r_n, r_{n+1}, r_{n+2})}{\sqrt{r_n^2 + r_{n+1}^2 + r_{n+2}^2}} \; . \qquad\qquad (5.3)$$

Any generator needs an initial value r_0, which we have set to $r_0 = 1234$ here. First, a short *Mathematica* program

```
        r0 = 1234
         a = 106
         c = 1283
         m = 6075
  rndm[r_] = Mod[a r+c,m]
  uniform = NestList[rndm,r0,m+1]/N[m]
```

provides us with all the random numbers, normalized to the unit interval, from which we obtain the unit vectors described in (5.3) via

```
    triple = Table[Take[uniform,{n,n+2}],{n,m}]
unitvectors = Map[(#/Sqrt[#.#])&,triple]
```

The following command turns the vectors e_n into graphics objects and displays them on the screen:

```
Show[Graphics3D[Map[Point,unitvectors]],
    ViewPoint -> {2,3,2}]
```

The vector $\{2,3,2\}$ points towards the observer. The left panel of Fig. 5.1 shows the result. If the points e_n were really distributed randomly, they would have to cover this segment of the surface evenly, except for a geometry factor resulting from the projection of the cube onto the sphere. Instead, one can see clear structures that result from the correlation between three consecutive numbers r_n. Indeed, it can be shown that the points $u_n = (r_n, r_{n+1}, r_{n+2})/m$ lie on a regular lattice, like the building blocks of a crystal. Starting from one point of the lattice, one can find its nearest neighbors and use them to determine the primitive lattice vectors. This lattice structure becomes visible when looking at the random points from a sufficiently large distance and a suitable direction. The right panel of Fig. 5.1 is the result of

```
Show[Graphics3D[Map[Point, triple]],
    ViewPoint -> {-325., 2000., -525.}]
```

It is obvious that a total of 14 planes is sufficient to take up all points. Still, this value has to be considered a relatively good result, since it has to be compared to the completely regular cubic structure that leads to $m^{1/3} = 6075^{1/3} \simeq 18$ planes. Thus, for a period of $m = 6075$, the vectors cannot be distributed more evenly than on 18 planes. A bad choice of a and c means that fewer planes are enough to take up all the points.

In Fig. 5.2, the random numbers are plotted in the same manner as in the left panel of Fig. 5.1, only this time they were generated by *Mathematica*'s generator `Random[]`. Now no more structures can be noticed on the surface of the sphere. This, however, does not mean that there are no other correlations between the random numbers. If the sequence is displayed in a suitable way, even higher correlations or interdependencies between well-separated numbers might become visible to the eye.

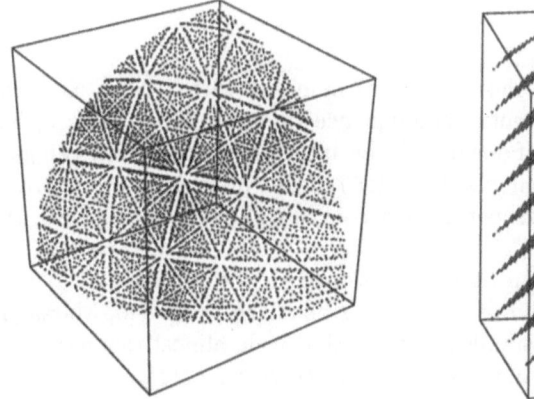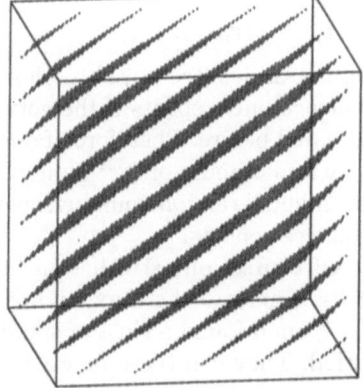

Fig. 5.1. Triples of pseudo-random numbers. Projection onto the surface of the unit sphere (*left*). The same points in the unit cube, viewed from a suitable direction (*right*)

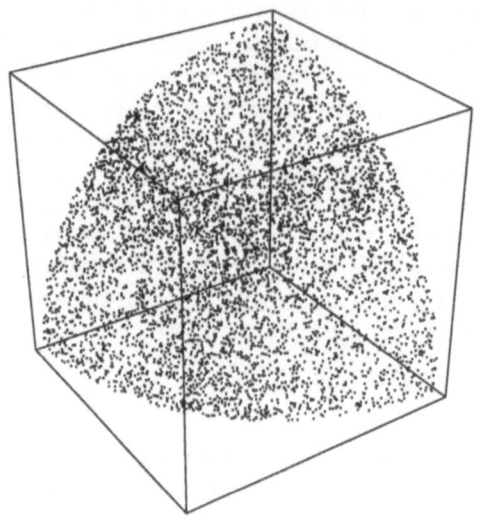

Fig. 5.2. The same plot as in the left panel of Fig. 5.1, except that the vectors were generated by *Mathematica*'s random number generator

There are various tricks one can use to obtain the longest possible periods and good random numbers. One can nest several different generators or use different types of functions f. Also, one can generate bit sequences by combining bits at a fixed distance via a Boolean operation. This operation generates a new bit, after which the sequence is shifted by one position for the next operation. It is known, though, that such shift register generators yield correlations in the bit sequences as well.

Using simple functions, how can one generate periods longer than m from m different numbers at all? To achieve this one can extend (5.1), for example by calculating r_n not only from one but from two or more previously generated numbers r_j:

$$r_n = f\left(r_{n-t}, r_{n-s}\right) . \tag{5.4}$$

Here, t and s are suitably chosen natural numbers with $t > s$. Now the sequence only repeats if the entire subsequence $\{r_{n-t}, r_{n-t+1}, \ldots, r_{n-1}\}$ has appeared before. Therefore, if one uses 32-bit integers, the generator can generate a maximum of 2^{32t} numbers. Even for $t = 2$, a program that generates one million random numbers per second can run $584\,942$ years before the sequence repeats.

Of course, one still has to find a function f in the iteration (5.4) that really generates the maximum period 2^{32t}. About ten years ago, Marsaglia and Zaman found a class of simple functions that yield almost the maximum period. They call these functions *subtract-with-borrow generators*:

$$r_n = (r_{n-s} - r_{n-t} - c_{n-1}) \bmod m . \tag{5.5}$$

Here, c_n is a bit that is set to $c_n = 0$ for the next step if $r_{n-s} - r_{n-t} - c_{n-1}$ is positive, and to $c_n = 0$ otherwise. The first c value needed can be set to 0 or 1 randomly, just as the values $r_1, r_2, \ldots, r_t$ can be initialized by suitable random numbers. Using arguments from number theory, the authors specify values of s, t, and m for which the period can be calculated. For example, the generator

$$r_n = (r_{n-2} - r_{n-3} - c_{n-1}) \bmod \left(2^{32} - 18\right) \tag{5.6}$$

has a period $(m^3 - m^2)/3 \simeq 2^{95}$. Moreover, this generator is excellently suited for the language C. In modern computers, the operation modulo 2^{32} is automatically built into the arithmetics of 32-bit integers. Since C allows unsigned integers (unsigned long int), (5.6) can be programmed very easily.

Marsaglia and Zaman combine this generator with

$$r_n = (69\,069\,r_{n-1} + 1\,013\,904\,243) \bmod 2^{32} \,, \tag{5.7}$$

which has a period 2^{32}. Equations (5.6) and (5.7) can be formulated easily in a C program which calculates the average of $N(= 1\,000\,000)$ of the resulting pseudo-random numbers:

```
#define N 1000000
typedef unsigned long int unlong;
unlong x=521288629, y=362436069, z=16163801,
       c=1, n=1131199209;

unlong mzran()
{ unlong s;
     if(y>x+c) {s=y-(x+c); c=0;}
          else        {s=y-(x+c)-18; c=1;}
     x=y; y=z; z=s; n=69069*n+1013904243;
     return (z+n);
}

main()
{
     double r=0.;
     long i;
     for (i=0;i<N; i++)
         r+=mzran()/4294967296.;
     printf ("r= %lf \n",r/(double)N);
}
```

x, y, z, and n are set to initial values which the user can change arbitrarily. The auxiliary bit c has to be set to c = y > z, i.e., in C to 1 (true) or 0 (false). Since this program was written for a computer that performs the operation modulo 2^{32} automatically, $\bmod(2^{32} - 18)$ reduces to subtracting 18 from the difference $(r_{n-2} - r_{n-3} - c_{n-1})$. Of course, this only happens if this difference is negative, as the modulo operation does not occur otherwise. The next random number is the sum of the two generators (5.6) and (5.7).

Because the periods of these two generators are relatively prime, the length of the combined generator is the product of the two individual lengths, about 2^{127}. Therefore, this generator can run for about 10^{24} years. Since, in addition to this, the sequence of numbers has passed many statistical tests, one can assume for many applications that these numbers behave like "genuine" random numbers.

Still, we have to warn the reader: The pseudo-random numbers above, as well as any other sequence, are the result of a deterministic algorithm, which can be analyzed by means of number theory. Consequently, it cannot be excluded that there will be correlations that interfere with statistical applications. Each problem reacts differently to different correlations. Therefore, even after several decades of experience with random numbers, experts can only give the advice to use different generators and compare the results. If possible, one should look for a similar problem that can be solved exactly and compare the numerical results to the exact ones. This provides a certain control over the quality of the random numbers. As future computer simulations will become more extensive and precise, the quality of the random numbers must not be neglected.

The examples in our textbook are only intended as demonstrations, not for precise quantitative analysis. In the following sections we will therefore use only the generators provided within the programming languages, **rand()** or **Random[]** respectively.

Exercises

1. Program the two random number generators

   $(I) \qquad r_n = (r_{n-2} - r_{n-5}) \bmod 10 \,,$

   $(II) \qquad r_n = (r_{n-2} - r_{n-5} - c_{n-1}) \bmod 10 \,,$

   where we set $c_n = 1$ or $c_n = 0$ if $(r_{n-2} - r_{n-5} - c_{n-1}) \leq 0$ or > 0 respectively. Both generators produce integers $0, 1, \ldots, 9$ and repeat the sequence if the last five numbers have appeared before in the same order. Consequently, both have to repeat the sequence generated after at most 10^5 steps.

   Start with different sets of five initial digits and calculate the actual periods of the sequences generated.

   Hint: The period of a generator can be calculated, for example, by generating two sequences in parallel. For the second sequence, one generates two steps at a time and then one waits for this sequence to overtake the first one.

2. Generate the sum s of N random numbers r distributed uniformly in the unit interval. The mean value of the sum is $m = 0.5N$ and the mean square deviation, or variance, is

$$\sigma^2 = N\left(\langle r^2 \rangle - \langle r \rangle^2\right) = N\left[\int_0^1 r^2 \, dr - \left(\int_0^1 r \, dr\right)^2\right] = \frac{N}{12} \,.$$

Repeat the experiment many times and plot the frequency distribution of the sums in a histogram. Compare the result to a Gaussian distribution with mean m and variance σ^2,

$$P(s) = \frac{1}{\sqrt{2\pi}\sigma} \exp\left[\frac{-(s-m)^2}{2\sigma^2}\right].$$

Verify the agreement for $N = 10$, 100, and 1000.

Literature

Binder K., Heermann D.W. (1992) Monte Carlo Simulation in Statistical Physics: An Introduction. Springer, Berlin, Heidelberg, New York

Gould H., Tobochnik J. (1996) An Introduction to Computer Simulation Methods: Applications to Physical Systems. Addison-Wesley, Reading, MA

Knuth D.E. (1997) The Art of Computer Programming, Vols. I, II, and III. Addison-Wesley, Reading, MA

Marsaglia G., Zaman A. (1994) Some Portable Very-long Period Random Number Generators. Computers in Physics 8:117

Press W.H., Teukolsky S.A., Vetterling W.T., Flannery B.P. (1992) Numerical Recipes in C: The Art of Scientific Computing. Cambridge University Press, Cambridge, New York

5.2 Fractal Aggregates

In Sect. 3.3 we have constructed objects that are more than a line but less than an area. These objects are characterized by a fractional dimension and are called fractals. They are self-similar, a part looks similar to the whole object, and they can be generated according to a simple deterministic rule.

Do such objects exist in nature? Indeed, many structures in our environment can be characterized quantitatively with the help of fractal dimensions. Coastlines, mountain ranges, courses of rivers, blood vessels, nerve cells, surfaces of leaves, variations of stock prices, and many other phenomena can all be described by power laws of the kind "the mass M increases as the length L to the power D." Natural structures, however, are not as strictly regular as the Sierpinski gasket from Sect. 3.3 but are the results of random processes.

Aggregates are a simple example of such fractals arising from the interplay of chance and regularity. If particles diffuse towards a nucleus and attach to it, the result is a loose grainy structure with a fractional dimension between 2 and 3. Using a simple computer model that was proposed and investigated in 1981 by Witten and Sander, we want to demonstrate how such fractal objects can be generated with little effort on the computer.

Physics

In the following, we describe a model of a growth process, in which the undirected diffusion of particles before their attachment plays the decisive role. The deposition of material from an electrolytic solution at very low voltage can serve as an example for this. One only has to make sure that the random motion of the diffusing ions is more important than their drift due to the electric field. Then the deposition of material on the electrode does not lead to a compact object; instead, a filigree-like cluster emerges, which we can describe by a fractal dimension D. The value of D is about 5/6 of the dimension of the space in which the growth process takes place.

The process is called *diffusion limited aggregation* (DLA). Though DLA is easily simulated and analyzed on the computer, to our knowledge there is no analytic theory for it to date. By contrast, there is a well-developed mathematical theory for the diffusion problem, i.e., the random motion of a particle. We want to briefly elaborate on a few statements and results which we will need in the algorithm section.

We consider a random walk on a square lattice or, more generally, a d-dimensional cubic lattice. We designate the lattice vectors by x and the vectors connecting a lattice site to its $2d$ nearest neighbors by $\Delta x_i, i = 1, 2, \ldots, 2d$. The motion of our random walker, which is at the lattice site x at time t, is determined by the following instructions: randomly select one of the nearest neighbor sites $x + \Delta x_i$ and jump there during the next time step Δt. If we now put many particles onto the lattice, all of which jump according to these same random rules, and designate the fraction that will be at the site x at time t by $u(x, t)$, we obtain the following balance equation:

$$u\left(x, t + \Delta t\right) - u\left(x, t\right) = \frac{1}{2d} \sum_{i=1}^{2d} \left(u\left(x + \Delta x_i, t\right) - u\left(x, t\right)\right) . \tag{5.8}$$

We have subtracted the term $u(x, t)$ on both sides to make the fact that this is just the discretized version of the diffusion equation more obvious. To see this, we divide (5.8) both by a suitable volume, to obtain a density, and by Δt and let all small quantities, in particular all Δx_i, go to 0 in proportion to $\sqrt{\Delta t}$. This leads to the equation

$$\frac{\partial u}{\partial t} = \eta \nabla^2 u \tag{5.9}$$

for the probability density, with a diffusion constant η.

We want to discuss two characteristic solutions of this diffusion equation. The first is the solution of an initial-value problem which contains information about the manner in which a random walk spreads out on average and the velocity with which that happens. The second is a stationary solution from which we can obtain an estimate of the fractal dimension D of the DLA cluster to which the diffusing particles attach.

Equation (5.9) with the initial condition $u(x, t_0) = \delta(x - x_0)$ and the boundary condition $u \to 0$ for $|x| \to \infty$ can be solved by a Fourier transformation. For $t \geq t_0$ one obtains

$$u(x, t) = [4\pi\eta(t - t_0)]^{-d/2} \exp\left(-\frac{|x - x_0|^2}{4\eta(t - t_0)}\right) . \tag{5.10}$$

From this we see that, if the jumping particle is at the point x_0 at the time t_0, then the probability of finding it at x at a later time t only depends on the distance $r = |x - x_0|$, not on the direction. This does not come as a big surprise, as it is just a reflection of the isotropy of the diffusion process which we have put into the equation by treating all directions identically. The mean value $\langle (x - x_0)^2 \rangle$ is

$$\left\langle (x - x_0)^2 \right\rangle = 2d\eta(t - t_0) , \tag{5.11}$$

which means that the average size of the random path increases as $\sqrt{t - t_0}$. Except for the proportionality factor, this law is independent of the dimension d of the space in which the random motion takes place.

The model of Witten and Sander describes a diffusion limited growth process which proceeds according to the following rules. One starts with a minimal cluster, usually a single particle at the origin of the coordinate system. Then, another particle starts diffusing freely at a great distance and at a randomly selected direction, until it hits the nucleus at the origin and attaches to it. Then the next particle starts far out, again at a randomly selected direction, and so on. When averaged over many realizations of this experiment, the central cluster will grow in a radially symmetric way. For any given realization, we designate the maximum radial size of the cluster by $R_{\mathrm{max}}(t)$ and determine its fractal dimension D from the relationship between $R_{\mathrm{max}}(t)$ and its mass $M(t)$ which is given by the number of particles attached by the time t:

$$M \propto R_{\mathrm{max}}^D . \tag{5.12}$$

If we want to describe this model by a probability density $u(x, t)$ of the diffusing particles, we have to take into account the boundary condition that the density vanishes at the newly created cluster. Additional particles are deposited on the cluster with a probability that is proportional to the particle current. We can, however, approximate this by assuming that all particles that get to within a certain capture radius R_{c} are removed by absorption. But according to Witten and Sander, R_{c} is proportional to R_{max}, and we do not commit an error in the following dimensional consideration if we replace R_{c} by R_{max}. Moreover, we can assume that the time that characterizes the diffusion process, e.g., the time it takes a diffusing particle to move from one site to the next, is much smaller than the time that characterizes the growth of the cluster. In other words, we are looking for a radially symmetric, time-independent solution of the diffusion equation (5.9) with the boundary condition $u(R_{\max}) = 0$. The equation

$$\Delta u = \frac{1}{r^{d-1}} \frac{\partial}{\partial r} r^{d-1} \frac{\partial}{\partial r} u = 0 , \tag{5.13}$$

however, is easily integrated. For $d > 2$, we get the solution

$$u(r) = u_0 \left(1 - \left(\frac{R_{\max}}{r} \right)^{d-2} \right) , \tag{5.14}$$

and u is proportional to $\ln(r/R_{\max})$ for $d = 2$. In any case, the radial component of the current density, $j_r = -\eta \partial u / \partial r$, is proportional to $R_{\max}^{d-2}/r^{d-1}$. By integrating this over the surface we obtain the result that the total particle current J absorbed by the cluster is proportional to $R_{\max}^{d-2}$. On the other hand, J is proportional to the increase in mass, which leads to the following equation:

$$R_{\max}^{d-2} \propto J = \frac{dM}{dt} = \frac{dM}{dR_{\max}} \frac{dR_{\max}}{dt} \propto R_{\max}^{D-1} v . \tag{5.15}$$

This means that the velocity v with which the cluster size increases is proportional to $R_{\max}^{d-1-D}$. If we add the plausible assumption that v will in no case increase with increasing cluster size, then the exponent $d - 1 - D$ must not be positive, i.e., the relation $d - 1 \leq D$ must hold. On the other hand, the DLA cluster is embedded in d-dimensional space, so we obtain the relation

$$d - 1 \leq D \leq d \tag{5.16}$$

which has been confirmed for $d = 2, 3, \ldots, 8$ by numerical simulations.

Algorithm

We want to generate a DLA cluster on a two-dimensional square lattice. To accomplish this, we occupy the site at the center of the lattice, place another particle on the lattice at a distance R_s, and let it diffuse until it either attaches to the occupied site or exceeds a distance $R_k > R_s$ from that site. Consequently, the algorithm will be as follows:

1. Start with an empty square lattice and position a particle on the central lattice site.
2. Let the actual size of the aggregate be defined by the radius $R_{\max}$. Place a particle on a randomly selected site at a distance $R_s \gtrsim R_{\max}$ from the origin and let it jump to a randomly selected adjacent site.
3. If the particle reaches a site adjacent to the aggregate, it is added and $R_{\max}$ is increased if necessary. If a particle exceeds the distance R_k from the origin ($R_k > R_s > R_{\max}$), it is annihilated ("killed").
4. Iterate steps 2 and 3 and calculate $D = \ln N / \ln R_{\max}$, where N is the number of particles in the aggregate.

The sketch in Fig. 5.3 serves to illustrate the smallest circle around the aggregate, with radius R_{max}, the start circle, and the annihilation circle. However, the introduction of an annihilation circle skews the properties of the aggregate, as it neglects the fact that the diffusing particle can get back to the aggregate even from a distance $R > R_k$. Consequently, our algorithm describes the actual problem only in the limit $R_k \to \infty$. Large values of R_k, though, mean long computing times, and we do not expect any significant deviations for finite values of R_k. One can analytically calculate the probability that the particle will return to a given position on the start circle and use this to reset the particle in one step. This avoids the annihilation circle. For large aggregates, this is worth the effort, but we choose not to go into this here.

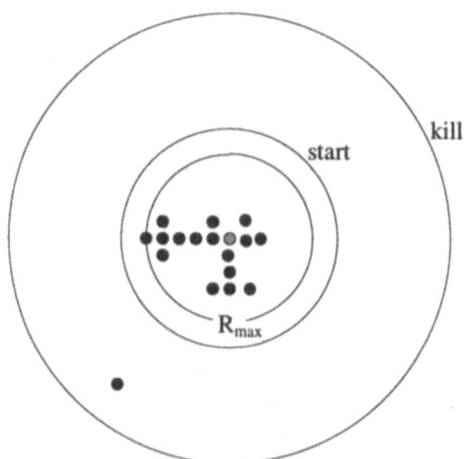

Fig. 5.3. The diffusing particle starts at the middle circle. If it leaves the outer circle, it is annihilated, and a new particle is put on the start circle

Moreover, the algorithm can be accelerated significantly by drawing a circle that just touches the aggregate around any particle that moves away from the cluster. Since, according to (5.10), the particle will get to each point on this circle with the same probability, one can randomly put it on the circle without simulating the time-consuming diffusion. This is repeated until the particle has reached a site adjacent to the aggregate or left the annihilation circle. This leads to the following algorithm:

1. Start with an empty square lattice and position a particle on the central lattice site.
2. Let the actual size of the aggregate be defined by the radius R_{max}. Place a particle on a randomly selected site at a distance $R_s \gtrsim R_{max}$ from the origin and let it jump to a randomly selected adjacent site.
3. If the particle reaches a site adjacent to the aggregate, it is added and R_{max} is increased if necessary. If the particle exceeds the distance $R_d > R_s \gtrsim R_{max}$ from the origin – the actual distance is denoted by R – it

is positioned on a randomly selected site on a circle around its current position with radius $(R - R_s)$. If the particle exceeds the distance $R_k > R_d > R_s \gtrsim R_{max}$, it is annihilated.

4. Iterate steps 2 and 3 and calculate $D = \ln N / \ln R_{max}$, where N is the number of particles in the aggregate.

Figure 5.4 is intended to clarify this algorithm once more. For a fast simulation of the DLA cluster, we use the programming language C. The square lattice is described by the two-dimensional array xf[rx][ry] whose elements can take the values 1 and 0. The assignment xf[rx][ry]=1 means that there is a cluster particle at the site with the coordinates rx and ry, whereas the sites with xf=0 are available for diffusion. For very large aggregates it would certainly make sense not to store the entire lattice, only the aggregate and boundary sites, but to avoid the bookkeeping that would be necessary in this case, we choose the easier way.

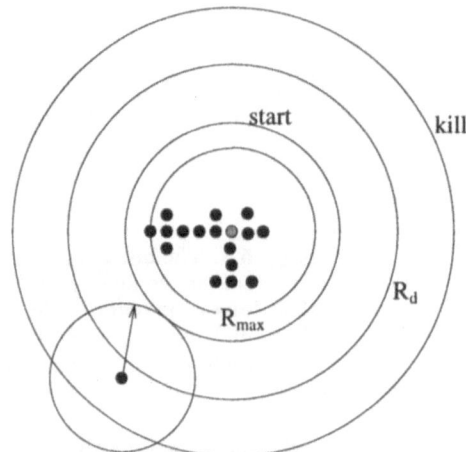

Fig. 5.4. The same situation as in Fig. 5.3, but outside the circle with radius R_d the particle covers the distance $R - R_s$ in one step

The start site on the circle of radius R_s is selected by the function occupy():

```
void occupy()
{
    double phi;
    phi=(double)rand()/RAND_MAX*2.*pi;
    rx=rs*sin(phi);
    ry=rs*cos(phi);
}
```

Jumping to one of the four adjacent sites is controlled via a randomly selected number 0, 1, 2, or 3:

```
void jump()
{
   int r;
   r=random(4);
   switch(r)
   {
     case 0: rx+=1;break;
     case 1: rx+=-1;break;
     case 2: ry+=1;break;
     case 3: ry+=-1;break;
   }
}
```

After each jump, the program has to check whether the particle is annihilated $(R > R_k)$, whether it has reached a site adjacent to the aggregate, or whether a short jump $(R < R_d)$ or a long "circle jump" $(R \geq R_d)$ needs to be performed in the next step. These four "ifs" in step 3 of the algorithm above are tested by the function check():

```
char check()
{
   double r,x,y;
   x=rx;
   y=ry;
   r=sqrt(x*x+y*y);
   if (r > rkill)   return 'k';
   if (r >= rd)     return 'c';
   if (xf[rx + 1 + lmax/2][ry + lmax/2] +
       xf[rx - 1 + lmax/2][ry + lmax/2] +
       xf[rx + lmax/2][ry + 1 + lmax/2] +
       xf[rx + lmax/2][ry - 1 + lmax/2] > 0)
     return 'a';
   else
     return 'j';
}
```

lmax is the size of the drawing window. The following function adds the particle to the aggregate:

```
void aggregate()
{
   double x,y;
   xf[rx+lmax/2][ry+lmax/2]=1;
   x=rx;y=ry;
   rmax= max(rmax,sqrt(x*x+y*y));
   if(rmax>lmax/2.-5.) {printf("\7");stop=1;}
   circle(4*rx+340,4*ry+240,2);
}
```

For the circle jump, a random vector of length $R - R_s$ is added to the position of the particle:

```
void circlejump()
{
  double r,x,y,phi;
  phi=(double)rand()/RAND_MAX*2.*pi;
  x=rx; y=ry; r=sqrt(x*x+y*y);
  rx+=(r-rs)*sin(phi);
  ry+=(r-rs)*cos(phi);
}
```

The particle's coordinates `rx` and `ry` and the aggregate array `xf[rx][ry]` are declared globally before the main routine `main()`, so all functions have access to them. With the five functions above, the essential part of the main routine can then be written as:

```
while(stop==0)
 {
  if(kbhit()) stop=1;
  switch(check())
   {
    case 'k':occupy();jump();break;
    case 'a':aggregate();occupy();jump();break;
    case 'j':jump();break;
    case 'c':circlejump();break;
   }
 }
```

Result

In Fig. 5.5 the result of the simulation can be seen. On the workstation, 3150 particles have attached themselves to a fractal aggregate within a minute.

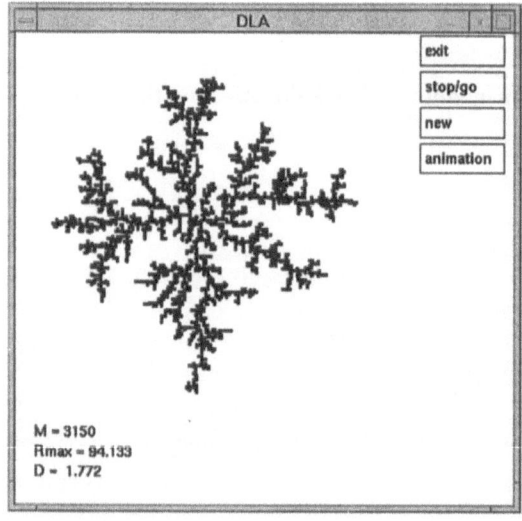

Fig. 5.5. DLA cluster on a square lattice

With a maximum extent of 94 lattice constants, one obtains the estimate $D = \ln 3150/\ln 94 \simeq 1.77$ for the fractal dimension.

Additionally the simulations show that, thanks to the introduction of the long jumps for $R > R_d$, the processing time does not increase significantly if the radius of the annihilation circle is enlarged.

Exercises

Improved dimensional analysis. In determining the fractal dimension we have, so far, neglected the fact that the cluster being generated is not complete yet. While the "arms" of the cluster continue to form, additional particles can still reach its interior regions. By arbitrarily stopping the simulation as soon as a certain maximum radius is reached, we neglect those particles. The actual cluster should be slightly denser, the fractal dimension slightly higher.

Calculate the mass of the cluster neglecting all particles whose distance from the origin is greater than r. Plot the mass as a function of r in a log–log diagram and attempt to interpret the behavior.

Varying the model parameters. It is interesting to observe in which way the form and fractal dimension of the cluster change if the original DLA model is varied slightly.

Introduce an additional parameter p_{stick} into the program dla.c, which represents the probability that a particle which arrives at a site adjacent to the cluster will actually stick to it.

If the particle is not attached, it shall continue the diffusion process until it hits the cluster again, at which point it will again have a probability p_{stick} of getting stuck. In continuing the diffusion process one has to make sure, of course, that the particle is prohibited from jumping to a site that is already occupied.

Literature

Gould H., Tobochnik J. (1996) An Introduction to Computer Simulation Methods: Applications to Physical Systems. Addison-Wesley, Reading, MA

Sander E., Sander L.M., Ziff R.M. (1994) Fractals and Fractal Correlations. Computers in Physics 8:420

5.3 Percolation

In an ideal crystal, all atoms are arranged on a lattice in a regular way. In this case we can use Fourier transformations to describe the periodic structures as well as the crystal excitations, e.g., the phonons. Since nature rarely consists of ideal crystals, however, solid state physicists have been trying for decades to understand materials with irregular structures as well.

There is a model for disordered materials, which permits a rather simple description: the *percolation model*. This model is intended to describe a porous material through which certain particles can diffuse (or *percolate*) if there are continuous paths through the pores. The geometry of the randomly generated pores is described by percolation theory. The model can be explained to any beginning student, it is easily programmed on a computer, yet it contains interesting physics related to phase transitions, critical phenomena, universality, and self-similar fractals.

In this section, we want to present and simulate the percolation model.

Physics

Let us assume that we have an alloy consisting of two kinds of atoms, which we label by A and B respectively. The concentration of A atoms is denoted by p, and both kinds of particles are assumed to be distributed randomly and to form a regular crystal lattice together. Let us further assume that the A atoms are magnetic, with a short-range magnetic interaction which is only effective if two A atoms are located next to one another. If there is no coupling with the nonmagnetic B atoms, the crystal can only form a magnetic order if there is an infinite network of connected A atoms. While, strictly speaking, phase transitions can only occur in infinite systems with a sufficient degree of interconnectedness, we will be able to see clear indications of a phase transition at a critical concentration of A atoms even for finite systems. The size of the structures of A atoms which are all connected with each other plays an important role in this process. Such a set of A atoms that are connected by their couplings is called a *cluster*.

For small concentrations p, all we get is a large number of small clusters but no overall magnetic ordering. As p increases, the average size of the clusters of A atoms will increase as well until, starting at a threshold value p_c, a cluster extends through the entire crystal. The value p_c is called the percolation threshold or critical concentration. For $p > p_c$, the infinite cluster coexists with many small ones; not until $p = 1$ are all sites occupied by A atoms and there is only one (infinite) cluster.

For the infinite crystal there is a single, well-defined value for the percolation threshold p_c. For example on the square lattice, with four nearest neighbors, one finds the numerical value $p_c = 0.59275 \pm 0.00003$; on the triangular lattice, with six nearest neighbors, $p_c = 1/2$ is known exactly. While p_c depends on the lattice type and on the range of the couplings, it does not depend on the specific realization of the sample, if the size of the crystal approaches infinity. In the finite crystal, on the other hand, there is of course only a certain probability, depending on p, that a cluster connecting two opposite sides exists.

Now one can ask many questions concerning the properties of the clusters and work out mathematical relations. What is the value of p_c for different lattice structures and cluster definitions? How does the average size of the finite clusters increase as p is increased up to the threshold p_c? How does the density of the infinite cluster increase above p_c? What does the distribution of cluster sizes look like? Does the infinite cluster have a structure at p_c?

The governing laws in the vicinity of the percolation threshold p_c are particularly interesting. For example, we define an average size $R(s)$ of a cluster consisting of s particles by

$$R^2 (s) = \frac{1}{s (s - 1)} \sum_{i \neq j} (r_i - r_j)^2 , \qquad (5.17)$$

where i and j number the particles of the cluster and r_i is the position of the ith atom on the lattice. Then the average size ξ of the finite clusters is defined as

$$\xi = \sqrt{\langle R^2(s) \rangle_{s < \infty}} . \qquad (5.18)$$

Here, $\langle \ldots \rangle_{s < \infty}$ designates the average over all finite clusters.

At the percolation threshold ξ diverges, and close to p_c we have

$$\xi \sim |p - p_c|^{-\nu} . \qquad (5.19)$$

One obtains this law for all percolation models. Surprisingly, the exponent ν is universal. This means that its value depends only on the spatial dimension. In the planar case, the value $\nu = 4/3$ is known exactly for different lattice types and ranges of the couplings, and even for the corresponding problem without a lattice, and for real two-dimensional alloys. For three dimensions, one numerically obtains $\nu \simeq 0.88$.

The probability $P(p)$ that an arbitrarily chosen A atom is part of the infinite cluster is zero for $p < p_c$ and increases continuously from $P(p_c) = 0$ to $P(1) = 1$. Close to p_c the rise is again described by a power law with a universal exponent β:

$$P (p) \sim (p - p_c)^{\beta} \qquad (p \gtrsim p_c) . \qquad (5.20)$$

In two dimensions we have $\beta = 5/36 \simeq 0.14$; this means that $P(p)$ rises very steeply near p_c.

While the density of the infinite cluster vanishes right at the percolation threshold, $P(p_c) \cdot p_c = 0$, it does exist there and has an interesting structure: it is a fractal. If we consider a square section of the lattice with sides of length L and designate the number of particles in this section that belong to the infinite cluster by $M(L)$, then at p_c we obtain

$$\langle M (L) \rangle \propto L^D , \qquad (5.21)$$

where $\langle \ldots \rangle$ designates the average over different sections. Thus the mass of the cluster increases as a power of the length, and according to Sect. 3.3 D

is its fractal dimension. In two dimensions, $D = 91/48 \simeq 1.89$. This shows that the critical percolation cluster is significantly more compact than the aggregate from the previous section, which was generated by diffusion.

So far, we have learned about three universal critical exponents ν, β, and D, but near p_c many more properties are described by power laws, with additional exponents. Singularities like these, however, are not independent of each other, but are connected by scaling laws. We only want to outline the idea here; for a deeper understanding, we refer the reader to textbooks about critical phase transitions and the theory of renormalization groups.

Let $M(p, L)$ be the average number of particles in a cluster that traverses a square with sides of length L. For each value of L, one obtains the number M as a function of p; this results in a whole set of curves. Scaling theory then tells us that close to the critical point this set of curves can, after suitable scaling, be represented by a single universal function f. The scales for the quantities M and L can be expressed by powers of either $p - p_c$ or the variable ξ defined in (5.19). In doing so, L is measured in units of ξ, and M in units of a power of ξ, e.g., ξ^x. This means

$$M(p, L) \sim \xi^x f\left(\frac{L}{\xi}\right) , \tag{5.22}$$

where f is a function which is not known initially and x is a critical exponent. If we set $L = k\xi$, with a constant k, then $x = D$ follows from (5.21). Thus, L is measured in units of ξ and M in units of ξ^D, hence the name '"scaling theory."

Further, from $L = k\xi$ it follows that

$$\frac{M(p, L)}{L^d} \sim \xi^{D-d} f(k) , \tag{5.23}$$

where L^d is the number of lattice sites in the d-dimensional cube with edges of length L. M/L^d, though, is the probability that a lattice site belongs to the percolation cluster. Therefore, for large values of L we get

$$\frac{M(p, L)}{L^d} \simeq pP(p) \sim (p - p_c)^\beta . \tag{5.24}$$

Equations (5.19), (5.23), and (5.24) lead to

$$(p - p_c)^\beta \sim (p - p_c)^{-\nu(D-d)} . \tag{5.25}$$

From this, one can conclude

$$\beta = (d - D)\nu . \tag{5.26}$$

This means that the three critical exponents are coupled by a scaling law. Indeed, the knowledge of two exponents is sufficient for calculating all others.

To begin with, the critical exponents are defined only for infinite lattices. With the computer, though, we can only populate finite lattices with particles. In finite systems, however, there is no phase transition, no sharply defined percolation threshold, and there are no divergences. How can one deduce the universal critical laws of the infinite lattice from the properties of a finite system?

The theory of *finite size scaling* answers this question. It is based on the scaling laws outlined above which, among other things, make statements about the dependence of the singularities on the lattice size L. Using the behavior of the correlation length described by (5.19) in the scaling relation (5.22) one obtains

$$M\left(p, L\right) \cdot |p - p_c|^{\nu D} \sim \tilde{f}\left((p - p_c)\, L^{1/\nu}\right) , \tag{5.27}$$

where $f(x) = \tilde{f}(x^{1/\nu})$. Consequently, the concentration p and the system size L are related to one another close to the critical point $p = p_c$, $L = \infty$.

The fractal dimension D is already obtainable from (5.21), since for $L \ll \xi$ the finite system looks just like the infinite one. Therefore, (5.21) yields $M(p_c, L) \sim L^D$. Thus one can numerically calculate the fractal dimension D from the increase of the number $M(p_c, L)$ of particles in the percolation cluster as a function of L. But how does one determine p_c and the exponent ν?

To do this, we establish a scaling law analogous to (5.27) for the probability $\pi(p, L)$ of finding a cluster connecting two opposite faces, which therefore percolates, in a sample of size L. Obviously,

$$\pi\left(p, \infty\right) = \begin{cases} 0 & \text{for } p < p_c , \\ 1 & \text{for } p > p_c . \end{cases} \tag{5.28}$$

For finite values of L, this step function is rounded off in the vicinity of p_c. Since π is constant for $p > p_c$ in the infinite system, the value of the corresponding scaling exponent is zero. Therefore, analogous to (5.27) for $M(p, L)$, the scaling equation for $\pi(p, L)$ is

$$\pi\left(p, L\right) = g\left((p - p_c)\, L^{1/\nu}\right) \tag{5.29}$$

with an unknown function g. If we now populate the sites of a lattice with atoms with a probability p, and then increase p, all clusters will grow. At a threshold value $p_c(L)$, a percolating cluster will appear.

The probability for $p_c(L)$ to lie in the interval $[p, p + dp]$ is given by the value of $(d\pi/dp)\, dp$ at the point p. The derivative $d\pi/dp$ exhibits a maximum that diverges for $L \to \infty$ and moves towards the percolation threshold p_c of the infinite lattice. Because of the large statistical fluctuations, the maximum of $d\pi/dp$ is not well determined numerically. It is better to determine the mean value and the fluctuations of $p_c(L)$. The mean value of the percolation threshold is given by

$$\langle p_c\left(L\right)\rangle = \int p \frac{d\pi}{dp} dp . \tag{5.30}$$

From (5.29), one obtains for this the scaling law

$$\langle p_c(L) \rangle = \int p L^{1/\nu} g' \left((p - p_c) L^{1/\nu} \right) dp \ . \tag{5.31}$$

With the substitution $z = (p - p_c) L^{1/\nu}$ and with $\int (d\pi/dp) dp = \pi(1, L) - \pi(0, L) = 1$, this leads to

$$\langle p_c(L) \rangle - p_c \sim L^{-1/\nu} \ . \tag{5.32}$$

If we determine the value $p_c(L)$ for as many systems of size L as possible, and repeat this for many values of L, which should be as large as possible, we can fit the mean value $\langle p_c(L) \rangle$ according to (5.32) and thereby determine p_c and ν. At the same time, we can calculate the mean value of the square of $p_c(L)$ and obtain from this a measure Δ of the width of $d\pi/dp$. For if we perform calculations analogous to (5.30) and (5.31) for

$$\Delta^2 = \left\langle \left(p_c(L) - \langle p_c(L) \rangle \right)^2 \right\rangle = \left\langle p_c(L)^2 \right\rangle - \langle p_c(L) \rangle^2$$

$$= \left\langle \left(p_c(L) - p_c \right)^2 \right\rangle - \langle p_c(L) - p_c \rangle^2 \ , \tag{5.33}$$

we obtain

$$\Delta \sim L^{-1/\nu} \ . \tag{5.34}$$

This means that the parameter ν can be obtained directly from the standard deviation of $p_c(L)$

Thus, by using the scaling laws, we are able to calculate the critical values for the infinite lattice from the properties of finite systems. This *finite size scaling* is not just limited to the percolation problem, but is valid for other kinds of phase transitions as well, e.g., for the magnetic transition of the Ising ferromagnet, a topic with which we will deal in Sect. 5.5. With this method, we have learned about an important tool for calculating universal, or model-independent, quantities.

Finally, we want to mention that, in addition to the *site percolation* discussed here, there is also the so-called *bond percolation*. In this case, not the *sites* of a lattice are populated with a probability p, but the *bonds* between adjacent sites. Such systems have a different percolation threshold, but they have the same universal critical properties as the model discussed.

Algorithm

On a lattice, the percolation structure is easily generated numerically. We choose a square lattice with sites (i, j), $i = 0, \ldots, L - 1$ and $j = 0, \ldots, L - 1$, and use uniformly distributed random numbers $r \in [0, 1]$. With this, the algorithm becomes:

1. Loop through all (i, j).
2. Draw a random number r.
3. If $r < p$, plot a point at the position (i, j).

The corresponding C program is:

```
double p = 0.59275;
int i, j, L = 500, pr;
pr = p*RAND_MAX;
for (i = 0; i < L; i++)
for (j = 0; j < L; j++)
    if (rand() < pr) putpixel(i,j,WHITE);
```

Figure 5.6 shows the result. One can see a pattern without any apparent structure. It is therefore surprising that one can extract so many mathematical regularities by quantitatively answering suitable questions. The analysis of the percolation structure, however, is not as easy to program as its generation. For example, we cannot easily answer the question of whether the structure in Fig. 5.6 percolates, i.e., whether there is a path connecting two opposite faces that only uses occupied sites. Even our visual sense is overtasked by this question.

We need an algorithm that can identify clusters of connected particles. A naive method would be to draw circles of increasing size around each occupied site and to mark all connected sites. This, however, takes a very large amount of processing time. A fast algorithm was developed in 1976 by Hoshen and Kopelman. It loops through the lattice only once and assigns cluster numbers to all occupied sites. Occasionally, though, parts of the same cluster are assigned different numbers. By using a bookkeeping trick, such conflicts are easily resolved at the end. The details of this algorithm are well described in the textbooks by Stauffer/Aharony and Gould/Tobochnik.

Here we want to pursue a different path, which was also suggested in 1976 by Leath. Clusters are to be generated directly by a growth process. To do

Fig. 5.6. Percolation structure on a square lattice at the percolation threshold

this, we start by occupying just the center of an empty lattice and then we define all adjacent sites as either occupied or free, with the probabilities p and $1 - p$ respectively. This process is iterated, converting for each occupied site the adjacent undefined sites to either free or occupied ones. When no undefined sites adjacent to the cluster are left, the growth stops; if the resulting cluster connects two opposite sides of the lattice, it is a percolation cluster. Many repetitions of this process yield a distribution of clusters that can be analyzed using the methods of finite size scaling. Since each particle generated gets the statistical weight p, and each free site gets the weight $(1 - p)$, and since the entire cluster is weighted with the product of all these factors, the growth process generates the clusters with the same probability as the previous algorithm.

Thus the growth algorithm for a cluster is:

1. Label all sites of a square lattice with sides of length L as "undefined" and occupy the center. Initially, therefore, the cluster consists of just one particle.
2. Pick an undefined site adjacent to the cluster.
3. Draw a uniformly distributed random number $r \in [0, 1]$ and occupy the site if $r < p$. Otherwise, label the site as "vacant."
4. Iterate 2 and 3 until no undefined sites adjacent to the cluster are left.

To make sure that the growth process is terminated in a well-defined manner at the edge of the lattice, we label the sites around the lattice boundary as "vacant." For item 2, we will generate a list, from which we can obtain the positions of the remaining undefined sites next to the cluster. Of course, this list has to be updated constantly during the growth process.

Since we want to simulate the largest possible lattices again and directly observe the growing cluster on the screen, we have written the algorithm in C. First, we label the three different cases with the numbers 0, 1, and 2:

```
#define NOTDEFINED 0
#define OCCUPIED 1
#define VACANT 2
```

For the lattice, which only needs to take these three values, we use a data type that uses little space in memory, namely `char array[L][L]`; additionally, we define a list that will hold the positions (i,j) of the undefined sites adjacent to the cluster via

```
struct {int x; int y;} list[PD];
```

This list, which initially contains only the four sites next to the center, is processed sequentially and receives new entries while this is taking place. Since the processed sites are no longer needed, this list can be written to and looped through periodically. Therefore, its length PD does not have to be equal to L^2; in our tests the choice PD $= 4L$ proved to be sufficient. The initialization is done as follows:

```
for(i=0;i<L;i++)
   array[0][i]=array[i][0]=array[i][L-1]=
                   array[L-1][i]=VACANT;
for(i=1;i<L-1;i++)
for(j=1;j<L-1;j++)
   array[i][j]=NOTDEFINED;
array[L/2][L/2]=OCCUPIED;
list[0].x=L/2+1; list[0].y=L/2;
list[1].x=L/2  ; list[1].y=L/2+1;
list[2].x=L/2-1; list[2].y=L/2;
list[3].x=L/2  ; list[3].y=L/2-1;
count=3;
putpixel(L/2, L/2, WHITE);
```

Here, count is a pure counting variable, whose value is incremented by 1 each time a new undefined boundary site of the cluster is generated. The main loop, item 4 above, is:

```
while(! done )
    {
      event();
      for(k=0; k<=count; k++)
        {
          i=list[k%PD].x; j=list[k%PD].y;
          definition(i,j);
        } /* for */
    } /* while */
```

Thus, the program loops through L and processes the list of still-undefined cluster boundary sites. The function definition(i,j) starts by checking whether the site (i,j) is already defined. This is necessary because the same boundary site may appear in the list more than once. If the site is not yet defined, step 3 is executed, and the site (i,j) is occupied if the random number r is smaller than the concentration p. If this newly occupied site has any undefined neighbors, their positions are entered in the list above, and the value of the variable count is increased accordingly.

In the program, this looks as follows:

```
void definition(int i, int j)
  {
    double r;
    if( array [i][j] != NOTDEFINED ) return;
    r = rand() / (double) RAND_MAX;
    if( r < p )
      {
        array[i][j] = OCCUPIED;
        putpixel (i, j, WHITE);
        if( array[i][j+1]==NOTDEFINED )
         {count++;list[count%PD].x=i;list[count%PD].y=j+1;}
        if( array[i][j-1]==NOTDEFINED )
         {count++;list[count%PD].x=i;list[count%PD].y=j-1;}
        if( array[i+1][j]==NOTDEFINED )
```

```
    {count++;list[count%PD].x=i+1;list[count%PD].y=j;}
    if( array[i-1][j]==NOTDEFINED )
    {count++;list[count%PD].x=i-1;list[count%PD].y=j;}
      }
    else array[i][j] = VACANT;
  }
```

The program still needs to be supplied with the declarations, the graphics initialization, and the input handling routine **event()**.

Results

To begin, we use the first algorithm for the generation of the entire percolation structure, and populate a square lattice of 500×500 sites with a concentration $p = 0.59275$. The occupied sites (A atoms) are marked in black, the lattice and the vacant sites are not shown. Figure 5.6 shows the result. One can only see noise, which does not give any indication of the mathematical laws that can be obtained from a quantitative analysis of this picture. We have selected the concentration p right at the percolation threshold, but neither a percolation cluster nor critical properties are recognizable. Only a suitable computer program can extract the clusters from this structure and quantitatively analyze them.

Figure 5.7 shows a cluster that was generated by the growth algorithm. The concentration p and the lattice size are the same as those in Fig. 5.6. Now one sees a percolating cluster, which exhibits structures on all length scales, from the lattice constant up to the size of the entire lattice. It is a fractal self-similar object. Of course, the program not only generates percolating clusters;

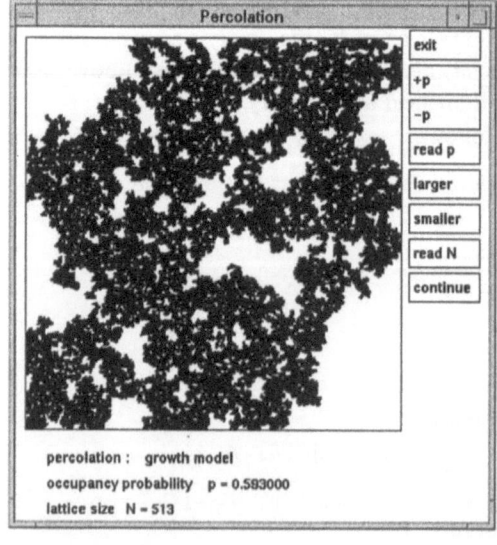

Fig. 5.7. Percolating cluster at the percolation threshold

very often the growth process stops before the cluster has reached the edge. For $p < p_c$ this occurs very frequently, whereas for $p > p_c$ percolating clusters are generated almost exclusively.

Exercise

Let s be the number of particles in a percolating cluster. In the infinitely large system, the average cluster mass $\langle s \rangle$ of the *finite* clusters diverges near the percolation threshold according to the power law

$$\langle s \rangle \sim |p - p_c|^{-\gamma} \ .$$

Calculate the mean value $\langle s \rangle$ for all numerically generated clusters that do not touch the edge. Plot these values as a function of the concentration p, and attempt to use the result to determine the critical concentration p_c. Calculate the maximum of this function for different values of the lattice size L. Try to calculate the critical exponent γ with the help of *finite size scaling*. In doing this, you may use the value of ν given above.

Literature

Gould H., Tobochnik J. (1996) An Introduction to Computer Simulation Methods: Applications to Physical Systems. Addison-Wesley, Reading, MA

Stauffer D., Aharony A. (1994) Introduction to Percolation Theory. Taylor & Francis, London, Bristol, PA

5.4 Polymer Chains

Polymers play an important role in chemistry as well as in biology and medicine. But physics, too, has long been interested in general mathematical laws concerning the properties of polymers. Even a single molecule which is made up of a long chain of identical units (*monomers*) has interesting properties. In order to be able to mathematically describe such chains of many thousands of monomers, one must attempt to model the essential properties and structures as simply as possible. One proven model for this is the randomly generated path, which is called the *random walk* in scientific literature. The path consists of many short line segments that are joined together in random directions.

In Sect. 3.3 we have already established that a random walk can be regarded as a model of a polymer molecule, all of whose configurations are equally probable in a heat reservoir. In that case, the mass of the chain increases as the square of its mean size. The mathematical theory for this is very well developed. This model neglects an important mechanism, however:

the chain cannot intersect itself. Random walks with such a restriction are appropriately called *self-avoiding walks* (SAWs). Although there are only approximate analytical calculations for SAWs, simulating them on a computer is relatively easy.

Physics

A random walk is most easily defined on a lattice. Since we are only interested in the global properties of very long polymers, in particular in their fractal dimension, we may assume that it does not make a difference if we consider a *random walk* with continuous step sizes and directions or one on a lattice. In Sect. 3.3 we have shown that an unrestricted random walk describes a tangled ball of a polymer molecule whose mass increases as the square of its mean size. For if N is the number of steps and R_N is the distance between the two ends of the walk, then

$$N = \frac{\langle R_N^2 \rangle}{a^2} . \tag{5.35}$$

Here $\langle \ldots \rangle$ is an average over all random walks and a is the length of the individual chain links. If we define a mean length by $L = \sqrt{\langle R_N^2 \rangle}$, we obtain

$$N \propto L^D \tag{5.36}$$

with the dimension $D = 2$. This result is valid not only in the plane, but in all spatial dimensions.

Now we consider random walks which may not intersect themselves. In real polymers it is the internal volume of the chains that prohibits this penetration. Obviously, such a self-avoiding walk will not be as bunched up on average as the unrestricted random walk; the mean end-to-end distance L will be larger. This means that the structure of the corresponding polymer will no longer be as compact and the dimension D should be less. Indeed, for large values of N, one finds that the SAW obeys a law of the form (5.36) again, but now with a fractal dimension D which depends on the spatial dimension d. As early as 1949, Flory derived a formula for D, using a kind of mean-field approximation:

$$D = \frac{d+2}{3} . \tag{5.37}$$

Subsequent investigations have shown that this formula is exact in $d = 1$, 2, and 4 spatial dimensions and that the numerical results for $d = 3$ yield a value for D which is only slightly higher. In $d = 4$ spatial dimensions, the value $D = 2$ agrees with the one from the random walk, indicating that, in this case, the prohibition of self-intersection is no longer relevant for the global properties. In all higher spatial dimensions the value of D remains 2. As with many other phase transitions, the critical exponents are described by mean-field theory starting with an upper critical dimension $d = 4$.

In fact, there are dilute polymer solutions in which, at high temperatures, an SAW law as in (5.36) and (5.37) has been observed experimentally. In the case of real molecule chains, there can also be attractive interactions between distant monomers of a single polymer molecule, causing the chain to collapse to a compact ball with $D = d$ at low temperatures. Of course, the situation becomes even more complex if many molecule chains interact with each other. Particularly in recent years, extensive numerical simulations have been able to contribute significantly to the understanding of the polymer dynamics of such a "spaghetti soup." Here, though, we want to concern ourselves with the simulation of the relatively simple SAW.

Algorithm

A random walk on the lattice is easily generated with a computer. For each step, one selects one of the adjacent sites via a random number and moves to that site. For a random walk that is not permitted to cross itself, the algorithm seems to be obvious as well: one randomly selects one of the adjacent sites not visited before and makes a step to this site. Surprisingly, though, the algorithm yields incorrect results: on average, the polymer becomes more compact than the correct statistical mean. Chains which touch in several places are generated more often than stretched chains with few contact points.

At this point we want to provide a brief explanation for this phenomenon. Consider a growing chain and let Z_i be the number of sites next to its head that have not been visited yet. Then $1/Z_i$ is the probability that a monomer is added at one of these sites, and the total probability W_N of generating a particular chain configuration with N links is

$$W_N = \prod_{i=1}^{N} \frac{1}{Z_i} . \tag{5.38}$$

One can see, therefore, that chains with many contact points (Z_i small) are assigned a high probability. This cannot be, however, since each allowed configuration must occur with the same probability in the correct statistical mean.

But (5.38) immediately points out a way to correct this unwanted preference for compact chains. One has to assign the statistical weight Z_i to each link and weight the properties of the polymer by this amount. This way each chain gets the same statistical weight

$$W_N \cdot \prod_{i=1}^{N} Z_i = 1 , \tag{5.39}$$

and consequently the correct value of the mean end-to-end distance L for this simulation is

$$L^2 = \langle R_N^2 \rangle = \frac{\sum_l R_{N,l}^2 \prod_{i=1}^{N} Z_{i,l}}{\sum_l \prod_{i=1}^{N} Z_{i,l}} \; , \qquad (5.40)$$

where l enumerates the configurations generated.

There is, however, an efficient method to directly generate polymer configurations with a constant probability. While the previous algorithm lets chains grow, this method, which is called *reptation*, works with chains of constant length N. In this process the polymer "slithers" across the lattice and thus constantly changes its form and direction.

Before we describe this reptation algorithm and its properties, we want to explain briefly what is to be understood by a configuration of the polymer and when two such configurations on the two-dimensional square lattice are considered to be different. For example, it makes sense to consider all those configurations that result from the possible translations of a given form of the polymer chain as being the same. To compare possibly different configurations, they can then be shifted in such a way that all begin at the same fixed point of the lattice. This already implies that the chain has a beginning and that we can consequently talk about the first monomer unit, the second, the third, etc.

There is one more symmetry of the square lattice that we should use for simplification, the symmetry with respect to 90° rotations. The direction of the first monomer might point to the north, west, south, or east.

If we ignore the direction of the first chain link, then the unique characterization of a configuration only requires the knowledge whether the next step after the first one is to the left, straight ahead, or to the right, and correspondingly for all subsequent steps. Ergo we can describe a configuration of a polymer consisting of N monomers by specifying a sequence of $N - 1$ directions whose possible values are *left*, *straight*, or *right*. In this context, it is important that we go through the sequence in a well-defined order. As we have already mentioned above, the polymer chain needs an orientation. To this end, we label its two ends as *head* and *tail* respectively and define, for example, that the orientation is to be specified from the tail end to the head.

But not every sequence of $N - 1$ directions represents an SAW; only those sequences that do not exhibit any self-intersections are allowed, a constraint that cannot be formulated locally. This is also the reason why there are so few analytical results concerning the statistics of SAWs up to now and why one has to rely on computer simulations instead.

The algorithm for the reptation of a polymer chain with N elements goes as follows:

1. Start with a configuration on the lattice, labeling the ends of the chain as *head* and *tail* respectively.
2. Remove the last monomer at the tail end and randomly select one of the sites adjacent to the head (three possibilities on the square lattice).

3. If this site is free, add a new head monomer there. If it is occupied, restore the old configuration, interchange the labels *head* and *tail*, and take the otherwise unmodified configuration into account in calculating averages.
4. Iterate steps 2 and 3.

It can be shown that this method generates all configurations that can possibly result from the starting configuration, with the same probability. To do this, we label the different reachable states with $l = 1, 2, 3, \ldots, \mathcal{N}$ and designate the occupancy probabilities that evolve after a suitably long time by $p_l(t)$. If in addition we know the transition probabilities $W(l \to k)$, with which in each time step a configuration l is transformed to another configuration k, we can write down the following equation for the time evolution of the $p_l(t)$:

$$\Delta p_l \equiv p_l(t+1) - p_l(t) = \sum_{k=1}^{\mathcal{N}} [W(k \to l)\, p_k - W(l \to k)\, p_l] \ . \tag{5.41}$$

This is a discretized version of the master equation. Because of the relation $\sum_k W(l \to k) = 1$, the summation of the second term can be executed. Since we are interested in the stationary distribution, i.e., in $\Delta p_l = 0$, we are looking for the solution of the following system of equations:

$$\sum_{k=1}^{\mathcal{N}} W(k \to l)\, p_k = p_l \ . \tag{5.42}$$

For some problems of this kind, the transition probabilities $W(l \to k)$ are constructed with the help of a stationary distribution w_k, in such a way that the equation

$$W(k \to l)\, w_k = W(l \to k)\, w_l \tag{5.43}$$

holds. In this case, one speaks of *detailed balance*, and one can see immediately from (5.41) that this leads to $p_k = w_k$ as a solution of the stationary master equation.

In our case, however, detailed balance is not valid, as is illustrated by the following example: The completely straight configuration has a probability of 1/3 of having a bend to the right at its head end after the next step, while the probability for this new state to return to the straight one is zero.

Instead of detailed balance, we can use the following relation for the reptation algorithm:

$$W(k \to l) = W(l_{\text{inv}} \to k_{\text{inv}}) \ , \tag{5.44}$$

where we designate by l_{inv} the state one obtains from l by interchanging *head* and *tail*. For $l = k_{\text{inv}}$ (5.44) is fulfilled trivially. This concerns all cases in which the algorithm requires an interchange of head and tail. For the other transitions, the following holds: If it is possible to remove a monomer at the tail end and add one at the head instead, then the inverse process, removing this new head monomer and putting it back at the tail end, is possible as well. From this we conclude that both sides of (5.44) are either equal to zero or not equal to zero. But if they are not equal to zero, they are both 1/3, as the addition of a new head monomer in a specific direction always happens with a probability 1/3, if it is at all possible. If we insert (5.44) into (5.42) we find, in a similar manner as above, that $p_l = \text{constant} = 1/\mathcal{N}$ is a solution since, of course, $\sum_k W(l_{\text{inv}} \to k_{\text{inv}}) = 1$ holds as well. This means that all obtainable configurations are generated with the same probability.

There are, however, entire classes of configurations which are unattainable by using the reptation algorithm if one uses the completely straight configuration, or an equivalent one, as the initial state. Figure 5.8 shows examples from two of these classes. We can assume, though, that the number of these configurations is small compared to the number of states in the main class. On the other hand, whether these configurations are relevant or not depends, of course, on the question asked. If we use the reptation algorithm to calculate the average end-to-end distance, we expect a value that is slightly too large because we have not taken the more compact states from the other classes into account.

We want to visualize the slithering polymer on the computer screen. To do so, we program the reptation algorithm on the square lattice in C. First, we declare the variable type `vector`, which defines structures with the spatial coordinates (x, y).

```
typedef struct { float x,y;} vector;
vector direction[3], polymer[NMAX];
```

`polymer[NMAX]` is an array of vectors containing all spatial coordinates of the polymer; for example, `polymer[5].y` specifies the y-coordinate of the fifth monomer. We start by defining the straight configuration for the polymer and then plot it by using the function `circle`:

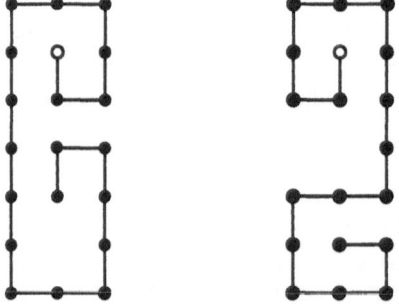

Fig. 5.8. Polymer configurations that are not reachable starting from a completely straight configuration. The open circle designates the head

```
for(i=0;i<N;i++)
  {
    polymer[i].x=i-N/2;
    polymer[i].y=0;
    circle(polymer[i]);
  }
```

Any other equivalent initial configuration is possible just as well. The labels
head and *tail* are initially assigned to the positions $(N-1)$ and 0 respectively,
and for each step these pointers are increased by 1 (modulo N).

The function choice() randomly chooses one of the three possible sites
adjacent to the head (item 2 above). The function intersection(c) checks
whether the chosen site is already occupied. If so, head and tail are inter-
changed, otherwise the new head is accepted and the tail element is overwritten. Thus, item 3 is realized in the program by:

```
while(!done)
  {
     event();
     c=choice();
     if(intersection(c))
            { head=tail; incr=-incr;
              tail=(head+incr+N)%N;
            }
       else accept(c);
  }
```

Rather than moving the polymer around in its memory location we use indices
for head and tail. The variable incr $= \pm 1$ specifies the direction of the
polymer with respect to the array polymer[N]. To avoid negative indices,
the number N is added before the modulo operation %.

As in all our C programs, the function event() intercepts keyboard or
mouse actions by the user and returns done=1 if appropriate, which terminates the while loop.

The function choice() returns an (x,y) vector pointing from the head to
one of the three adjacent sites. First the three possible directions are generated by calculating the difference $(r(head) - r(neck))$ – called direction[0]
– and then determining the two unit vectors perpendicular to this. For example, if the vector direction[0] is $(0,1)$, the variables direction[1] and
direction[2] contain the vectors $(1,0)$ and $(-1,0)$ respectively. Finally, using a uniformly distributed random number from the interval $[0,3)$, truncated
to an integer r, one of the three directions is randomly selected and returned:

```
vector choice()
  {
    int hm,r;
    hm=(head-incr+N)%N;
    direction[0].x=polymer[head].x-polymer[hm].x;
    direction[0].y=polymer[head].y-polymer[hm].y;
    direction[1].x=direction[0].y;
```

```
        direction[1].y=direction[0].x;
        direction[2].x=-direction[1].x;
        direction[2].y=-direction[1].y;
        r=random(3);
        return direction[r];
    }
```

The function `intersection(c)` returns the value 1 or 0 respectively, depending on whether the selected site is already occupied by the chain or not.

```
    int intersection(vector c)
    {
      int i;
      for(i=0;i<N;i++)
        if(c.x+polymer[head].x==polymer[i].x  &&
           c.y+polymer[head].y==polymer[i].y  &&
           i != tail)   return 1;
      return 0;
    }
```

If there is no overlap, the selected site is accepted as the new head and the indices are shifted by the value `incr`. On the screen, the circle at the location of the tail is painted over with the background color and a new circle for the head is added. The result is a serpentine motion on the lattice. All this is done by the following function:

```
    void accept (vector c)
    {
     int hp;
     void shift();
     setcolor(BLACK);
     circle(polymer[tail]);
     hp=(head+incr+N)%N;
     polymer[hp].x=c.x+polymer[head].x;
     polymer[hp].y=c.y+polymer[head].y;
     head=hp;
     tail=(head+incr+N)%N;
     setcolor(WHITE);
     circle(polymer[head]);
     if(abs(polymer[head].x)>2*N ||
        abs(polymer[head].y)>2*N) shift()
    }
```

To calculate the average end-to-end distance L, we have to take into account the length of the difference vector $r(head) - r(tail)$ after each step, including each reversal of direction, add all lengths, and finally divide by the number of reptation steps. In addition, we have defined a function `shift()`, which shifts the polymer to the center of the lattice whenever it is about to leave the window.

Results

Running our C program generates the polymer on the screen (Fig. 5.9). At the same time, the average end-to-end distance is printed, relative to the result $L = \sqrt{N}$ of the unrestricted random walk. It can be seen that, on average, the polymer is more stretched out than the random walk and that this relative average length increases with N, in agreement with the smaller fractal dimension $D = 4/3$. Indeed, just fitting the two values for $N = 50$ and $N = 100$ shows rather accurately that the ratio of the two lengths increases proportionally to $N^{0.25}$, as predicted by theory.

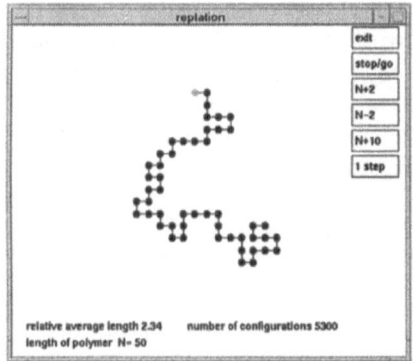 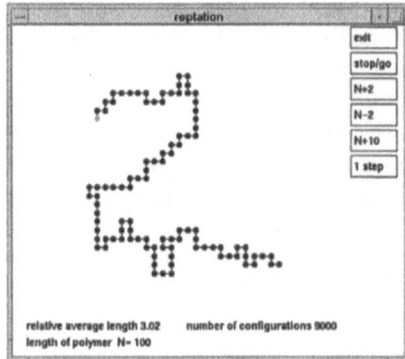

Fig. 5.9. Polymer chains of length $N = 50$ (*left*) and $N = 100$ (*right*) on the square lattice

Exercise

Program the algorithm mentioned at the beginning of the section by correctly taking into account the number Z_i of sites available in each step, as specified in (5.39) and (5.40). Compare this implementation to the reptation algorithm with respect to its speed and to the value of the mean end-to-end distance.

Literature

Binder K., Heermann D.W. (1992) Monte Carlo Simulation in Statistical Physics: An Introduction. Springer, Berlin, Heidelberg, New York

Gould H., Tobochnik J. (1996) An Introduction to Computer Simulation Methods: Applications to Physical Systems. Addison-Wesley, Reading, MA

5.5 The Ising Ferromagnet

If very many particles interact with each other, qualitatively new properties can be the result. For example, at low temperature or high pressure gases suddenly liquefy, liquids turn into solids, magnetic atoms align themselves to form a magnetic material, and metals suddenly lose their electric resistance. There is a large number of systems like these, which, in thermal equilibrium, change their macroscopic properties at a well-defined temperature. In mathematical modeling, these phase transitions only occur in infinitely large systems.

Is it possible to simulate thermal equilibrium and slowly cool down a system on the computer? Can the simulation of finite systems describe phase transitions? In this section, we want to investigate these kinds of questions, using a simple model as an example, the Ising ferromagnet on a square lattice. This model has universal critical properties at the phase transition, which are also found for many other models and measured in real materials. We let the microscopic magnets flicker on the screen and use keyboard inputs to heat up or cool down the magnet.

Physics

As early as the beginning of this century, the mathematical description of a system of particles in thermal equilibrium was known. In its canonical formulation, statistical mechanics traces back the thermal properties of a system to the calculation of a partition function Z:

$$Z = \sum_{\underline{S}} e^{-H(\underline{S})/k_\mathrm{B}T} . \tag{5.45}$$

Z is the sum over all possible multiparticle states $\underline{S}$, and each state is weighted according to its energy $H(\underline{S})$. T is the temperature and k_B the Boltzmann constant.

We want to describe the Ising model, a simple model for a magnetic system. To this end, we consider a square lattice whose lattice sites are occupied by spins with only two possible orientations. Additionally, there is an interaction between adjacent spins. Mathematically, this can be formulated as follows: For each lattice site $i = 1, \ldots, N$, there is a variable $S_i \in \{+1, -1\}$ such that a multiparticle state is identified by $\underline{S} = (S_1, S_2, \ldots, S_N)$. The calculation of the total energy $H(\underline{S})$ consists of a summation over all pairs of nearest neighbors $(i, j)_\mathrm{nn}$ as well as over all sites i:

$$H = -J \sum_{(i,j)_\mathrm{nn}} S_i S_j - h \sum_i S_i . \tag{5.46}$$

This means that parallel nearest-neighbor spins have a pair interaction energy $-J < 0$, while the term proportional to h describes an external magnetic field in which every spin that is parallel to the field has the energy $-h < 0$.

Thus the problem is well-defined: we "only" have to evaluate the sum (5.45). We want to attempt this using a computer. Let us assume we have one hundred hours of processing time available on a machine which needs about 10^{-6} s for each step of the calculation. This assumes that we have a fast computer, since one computation step, for example the evaluation of the exponential function, takes many clock cycles. Altogether, we can then execute 3.6×10^{11} computations.

How many steps does it take to evaluate (5.45) for N spins? Since each spin S_i has two possible orientations, there is a total of 2^N different spin configurations. For each of these states, the energy H has to be calculated according to (5.46), which takes $2N$ computation steps. This leads to

$$2N2^N = 3.6 \cdot 10^{11} \tag{5.47}$$

with the solution $N \simeq 32$. The result is discouraging. Even with today's supercomputers, we can only calculate a tiny magnet of size $3 \times 3 \times 3$, whereas we want to describe a real material with $10^7 \times 10^7 \times 10^7$ spins!

By using transfer matrix methods it is possible to calculate the partition function Z for rods of the size $4 \times 4 \times \infty$. But this is still very little if we want to observe phase transitions.

One could now get the idea to just take a few randomly generated states $\underline{S} = (S_1, S_2, S_3, \ldots, S_N)$ into account when calculating the partition function (5.45). But this does not work, because, according to the central limit theorem, randomly generated states have an energy $H \simeq 0 + \mathcal{O}(\sqrt{N})$ for large N. The physically important states, however, have an energy of the order $\mathcal{O}(N)$, which means that they are not generated at all with the method above.

Ever since the era of the first computers, physicists have investigated how the physically relevant states can be generated. To do this, one starts with a configuration $\underline{S}(t = 0)$ and generates from it a sequence $\underline{S}(1), \underline{S}(2), \ldots, \underline{S}(t)$ which relaxes into thermal equilibrium. To this end, a transition probability $W(\underline{S} \to \underline{S}')$ is defined, which – cf. the previous section – obeys the following detailed balance:

$$W\left(\underline{S} \to \underline{S}'\right) e^{-H(\underline{S})/k_B T} = W\left(\underline{S}' \to \underline{S}\right) e^{-H(\underline{S}')/k_B T} . \tag{5.48}$$

This equation can be understood by the following considerations: In thermal equilibrium,

$$P\left(\underline{S}\right) = \frac{1}{Z} e^{-H(\underline{S})/k_B T} \tag{5.49}$$

is the probability of finding a given state $\underline{S}$. The detailed balance requirement (5.48) has the effect that the thermal equilibrium $P(\underline{S})$ will be a stationary state for the dynamics defined by W. The occupancy probability $P(\underline{S})$ does not change with time. As explained briefly in Sect. 5.4, this can be formulated in a mathematically precise way and proven with a master equation.

In the algorithm part below we will define a function $W(\underline{S} \to \underline{S}')$ which obeys (5.48) and is easily programmable. In principle, (5.48) still leaves us a lot of freedom in the choice of W; we just have to make sure that W can cover the entire configuration space (= set of states $\underline{S}$). Two of the most common algorithms which fulfill both conditions are known as the *Metropolis algorithm* and the *heat-bath algorithm*.

The stochastic process, defined by W, leads to a sequence of physically relevant states $\underline{S}(0), \underline{S}(1), \ldots, \underline{S}(t)$ which, except for fluctuations, yield for example the correct energy and magnetization. With these states $\underline{S}(t)$ averages of quantities $A(\underline{S})$ can be calculated:

$$\langle A \rangle_{t_0, t_1} = \frac{1}{t_1} \sum_{t=t_0}^{t_0+t_1} A\left(\underline{S}(t)\right) . \tag{5.50}$$

In the limit $t_0 \to \infty$, $t_1 \to \infty$, this time average agrees with the statistical average

$$\langle A \rangle = \sum_{\underline{S}} P\left(\underline{S}\right) A\left(\underline{S}\right) . \tag{5.51}$$

In practice, however, one can only work with finite values for t_0 and t_1. In doing so, t_0 has to be large enough for the system to relax into thermal equilibrium from its potentially unphysical initial state $\underline{S}(0)$, and t_1 must be so large that the statistical fluctuations of $\langle A \rangle_{t_0, t_1}$ are small. Both requirements depend on the model, its parameters, and the system size.

With today's computers one can reach approximately the following values:

$$N \simeq 10^6 \ldots 10^{12} , \quad t_0 \simeq t_1 \simeq \left(10^4 \ldots 10^6\right) N . \tag{5.52}$$

Therefore, only about 10^{10} to 10^{15} configurations $\underline{S}(t)$ are generated, while the model has 2^N, or $10^{300\,000}$ to $10^{300\,000\,000\,000}$, different states $\underline{S}$. Whichever algorithm one uses, it can only generate a tiny fraction of all the possible configurations. But it has to generate the "correct" states, i.e., the typical or physically relevant ones.

Before we get to the issue of programming the stochastic process described by (5.48), we want to briefly discuss the properties of the Ising ferromagnet and their dependence on the system size N. The infinite system ($N \to \infty$) in two or more spatial dimensions has a phase transition in the absence of an external magnetic field ($h = 0$). Below a critical temperature T_c, the spins align themselves parallel to each other in the thermal average and generate a macroscopic magnetization

$$M(T) = \frac{1}{N} \sum_i \langle S_i \rangle . \tag{5.53}$$

M can be positive or negative; the system selects one direction through random fluctuations while cooling down. Since for $h = 0$ the energy is symmetric in M, $H(\underline{S}) = H(-\underline{S})$, this process is called *spontaneous symmetry breaking*. An external magnetic field $h \neq 0$ cancels this symmetry and thus destroys the phase transition.

The phase transition is critical, which means that, as the temperature increases, the magnetization $M(T)$ changes continuously towards $M(T) = 0$ for $T \geq T_c$. As in the case of percolation, see Sect. 5.3, there is a diverging correlation length $\xi(T)$ which is defined by the decay of the spin correlations for large distances r:

$$\langle S_i S_j \rangle \sim \exp\left(\frac{-r}{\xi}\right) \quad \text{where} \quad r = |r_i - r_j| . \tag{5.54}$$

In the vicinity of the phase transition, $T \simeq T_c, h \simeq 0$, there are again scaling laws and power singularities with universal critical exponents. Near T_c, the following relations hold for the magnetization M, the correlation length ξ, the magnetic susceptibility χ, and the specific heat C:

$$M(T) \sim (T_c - T)^\beta \quad (T \nearrow T_c)$$
$$\chi(T) \sim |T_c - T|^{-\gamma}$$
$$\xi(T) \sim |T_c - T|^{-\nu}$$
$$C(T) \sim |T_c - T|^{-\alpha} . \tag{5.55}$$

The reasons for these universal properties are given by the mathematically intricate theory of the *renormalization group* from the seventies (Nobel prize for K.G. Wilson in 1982). The two-dimensional model (for $h = 0$) was solved in 1944 by Onsager. Consequently, the universal quantities are known exactly for $d = 2$; the Onsager solution gives

$$T_c = \frac{J}{k_B} \frac{2}{\ln(\sqrt{2}+1)} \simeq 2.269 \frac{J}{k_B} ,$$

$$\beta = \frac{1}{8} , \; \nu = 1 , \; \gamma = \frac{7}{4} , \; \alpha = 0 . \tag{5.56}$$

The specific heat C diverges as $\ln|T - T_c|$. Thus the value $\alpha = 0$ does not mean that the specific heat remains finite at T_c, it just means that it diverges more weakly than any power of $|T - T_c|$. The value of T_c is only valid for the Ising model on the square lattice, whereas the exponents are valid for any two-dimensional system whose critical phase transition has the same symmetries as the Ising model.

We have already seen in the case of percolation that a finite system allows us to draw conclusions regarding the behavior of the infinite system. This finite size scaling, which is also justified by renormalization theory, is valid for the Ising ferromagnet as well. A finite system has no phase transition to a magnetically ordered state, but the critical exponents can be determined from the dependence of χ and C on N.

Let L be the length of the side of a square with $N = L^2$ spins. Asymptotically close to T_c we have, analogous to (5.27):

$$\chi(T, L) = |T - T_c|^{-\gamma} f\left(|T - T_c| L^{1/\nu}\right) . \tag{5.57}$$

By setting $|T - T_c| L^{1/\nu} = \text{constant}$, one obtains

$$\chi(T_c, L) \sim L^{\gamma/\nu} .$$ (5.58)

Thus one can determine the exponent γ/ν from the divergence of the susceptibility as a function of the system size. Similarly, one finds the exponent $\ \ \ $ 'rom the divergence of the specific heat. From these two values all other exponents can be calculated by using the scaling laws. We get (d = spatial dimension 2, 3, or 4):

$$\nu = \frac{2}{d + \alpha/\nu} , \quad \gamma = \frac{\gamma}{\nu}\frac{2}{d + \alpha/\nu} , \quad \beta = \frac{d - \gamma/\nu}{d + \alpha/\nu} .$$ (5.59)

Algorithm

The stochastic process requires a transition probability $W(\underline{S} \to \underline{S}')$ which obeys the detailed balance (5.48). Here we want to use a simple method, the Metropolis algorithm. W shall only permit transitions in which at most one spin is changed. Let therefore

$$\underline{S} = (S_1, S_2, \ldots, S_i, \ldots, S_N)$$

and

$$\underline{S}' = (S_1, S_2, \ldots, -S_i, \ldots, S_N) .$$ (5.60)

Additionally, transitions to energetically favored states shall always happen ($W = 1$). With (5.48) and (5.49) we then obtain

$$W\left(\underline{S} \to \underline{S}'\right) = \begin{cases} 1 & \text{for } H\left(\underline{S}'\right) < H\left(\underline{S}\right) , \\ \exp\left(\frac{H(\underline{S}) - H(\underline{S}')}{k_B T}\right) & \text{otherwise} . \end{cases}$$ (5.61)

$\Delta E = H(\underline{S}) - H(\underline{S}')$ is easily calculated, for we have

$$H\left(\underline{S}\right) = -J S_i \sum_{j \in \mathcal{N}(i)} S_j - h S_i + remainder ,$$ (5.62)

where the sum over j is restricted to the nearest neighbors of i and *remainder* contains that part of H which does not depend on S_i. $\underline{S}'$ has the same spins $S'_j = S_j$ as $\underline{S}$, except $S'_i = -S_i$; consequently $H(\underline{S}')$ has the same remainder too. This leads to

$$\Delta E = -2 S_i \left(J \sum_{j \in \mathcal{N}(i)} S_j + h \right) = -2 S_i h_i .$$ (5.63)

The term h_i is called the internal field of the spin S_i. This finally enables us to formulate the Metropolis algorithm:

1. Choose an initial state $\underline{S}(0) = (S_1, \ldots, S_N)$.
2. Choose an i (randomly or sequentially) and calculate $\Delta E = -2 S_i h_i$.

3. If $\Delta E \geq 0$, then flip the spin, $S_i \to -S_i$. If $\Delta E < 0$, draw a uniformly distributed random number $r \in [0, 1]$. If $r < \exp(\Delta E / k_B T)$, flip the spin, $S_i \to -S_i$, otherwise take the old configuration into account once more.
4. Iterate 2 and 3.

This algorithm is not only valid for the Ising model, but generally for all models of statistical mechanics. We will also apply it to difficult optimization problems in the next section. Because this method uses random numbers, it is called *Monte Carlo simulation*.

The algorithm above is not the only possible way of fulfilling detailed balance and thus obtaining thermal equilibrium. In recent years in particular, there has been a multitude of new developments in connection with this method known for four decades. For example, one can simultaneously flip large, suitably chosen clusters of spins and thereby significantly accelerate the simulation, in particular near T_c. This method can be improved further by using a hierarchical superstructure, which is known from the *multigrid methods* of computer science. Another method uses the information about fluctuations of energy and magnetization to obtain properties for an entire temperature interval from one simulation. Also, one can perform simulations at a constant energy (microcanonically) or use a trick to surmount high energy barriers. The Metropolis algorithm, on the other hand, is simple and universal, which is why we want to program it here.

We are interested in making the system size N, the relaxation time t_0, and the measuring time t_1 as large as possible. Consequently, we need a fast programming language. In current research and for computer simulations of large systems Fortran is used predominantly, as the Fortran compilers of supercomputers possess the highest degree of optimization. But C programs run effectively on high-performance computers as well. In the following, we want to program the Monte Carlo simulation of the Ising ferromagnet in C.

The spins S_i are stored in a quadratic array s[x][y], x = 1, ..., L, y = 1, ..., L, and all are initialized to the value $S_i = 1$. One could take any arbitrary initial state, since the system always relaxes into thermal equilibrium. At low temperatures, though, other initial states may have long relaxation times. The diffusion of domain walls – an example of which can be seen in the results section – makes this obvious.

Boundary effects can have a strong influence on the properties of the magnet, particularly in small systems. For this reason, we use periodic boundary conditions, i.e., each spin on the boundary interacts with the spin on the opposite side. Thus the system does not really have a boundary but is topologically wrapped up to a torus (which can be visualized as a doughnut-shaped object). To save as much processing time as possible in the innermost loop, we do not make any effort at determining the boundary indices via if statements or modulo operations. Instead, we copy the boundary spins into four additional arrays s[0][i], s[L+1][i], s[i][0], and s[i][L+1], i = 1, ..., L.

The algorithm compares random numbers to $\exp(-2S_i h_i/k_B T)$, the Boltz-mann weight factor at the temperature T. With a little trick, we can avoid the repeated calculation of the exponential function. We demonstrate this trick for $h = 0$, but it works just as well for $h \neq 0$. We take advantage of the fact that the sum $S_i \sum_{j \in \mathcal{N}(i)} S_j$ over the nearest neighbors of S_i can only take the values -4, -2, 0, 2, and 4 on the square lattice. For negative values, S_i is always flipped. Consequently, there are only two different Boltzmann weight factors $\exp(-4/T)$ and $\exp(-8/T)$, where the temperature T is measured in units of the interaction energy J/k_B. Both numbers are calculated before the actual Monte Carlo steps, using the following function:

```
void setT(double t)
  {
    temp=t;
    bf[0]= 0.5*RAND_MAX;
    bf[1]= exp(-4./temp)*RAND_MAX;
    bf[2]= exp(-8./temp)*RAND_MAX;
  }
```

Of course the variable temp and the array bf[...] have to be declared be-fore the main program main(), otherwise the function setT() will not know these quantities. If the internal field is zero, we use the transition probability $W(\underline{S} \to \underline{S}') = W(\underline{S}' \to \underline{S}) = 1/2$ (instead of 1). This avoids oscillations which might otherwise occur for a sequential selection of the sites visited.

With the Boltzmann weight factors bf[...] thus defined, the innermost loop becomes

```
for(x=1;x<L+1;x++) for(y=1;y<L+1;y++)
  {
  e=s[x][y]*(s[x-1][y]+s[x+1][y]+s[x][y-1]+s[x][y+1]);
  if( e<0 || rand()<bf[e/2] )
    {
    s[x][y]=-s[x][y];
    v=2*(x*80+2*(y-1)+2);
    ch=(s[x][y]+1)*15;
    poke(VSEG,v,0xf00|ch);
    }
  }
for(x=1;x<L+1;x++)
  {
  s[0][x]   = s[L][x];
  s[L+1][x] = s[1][x];
  s[x][0]   = s[x][L];
  s[x][L+1] = s[x][1];
  }
```

In the last for loop, the auxiliary arrays are filled with the current boundary spins in order to fulfill the periodic boundary conditions. With the command poke we directly address the display memory on the PC in text mode. This is very fast; by comparison, putch() or a graphics command would slow down the program by about a factor two. This command, however, does not work on all machines.

The Monte Carlo simulation can be programmed with little effort. With short programs and fast computers one obtains results easily. Analyzing the data, on the other hand, is not easy. After all, one wants to deduce general laws regarding the properties of the infinite lattice in thermal equilibrium ($\hat{=}$ infinite relaxation times) from strongly fluctuating results for relatively small systems and – compared to the experiment – relatively short running times. Even with the theory of finite size scaling mentioned above, which of course is only valid asymptotically for $N \to \infty$, this remains a difficult and tedious task. Add to this that quite frequently hidden correlations in seemingly good random number generators have led to systematic errors.

Results

The PC program ISING simulates a 20×20 Ising ferromagnet on the square lattice and indicates the orientation of the spins on the screen. The system can be heated up or cooled down with a keystroke while running. The initial temperature $T = 2.269 J/k_B$ is the critical temperature of the infinite magnet, according to Onsager. In addition to the temperature T and the number of Monte Carlo steps per spin (MCS), the program prints the time it takes the computer to execute one Monte Carlo step, about 4×10^{-5} s on our PC. Thus, even on the PC, a step is so fast that the spins in the simulated heat bath flicker. Figure 5.10 shows a system of size 100×100 whose behavior was simulated with the corresponding program on the workstation. At T_c the picture looks self-similar, i.e., clusters of parallel spins show up at all scales. At the critical point, the correlation length ξ is the same as the lattice size L, and for $a \ll r = |r_i - r_j| \ll L$ the correlations $\langle S_i S_j \rangle$ decay as a power of r (a = lattice constant). At $T = 0.8T_c$, the system is magnetized,

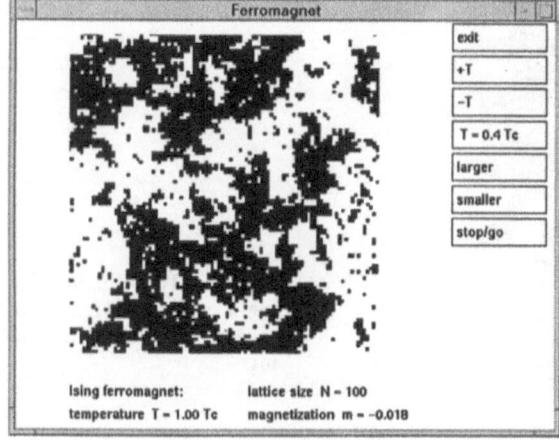

Ising ferromagnet: lattice size N = 100
temperature T = 1.00 Tc   magnetization m = –0.018

Fig. 5.10. State of the ferromagnet in thermal equilibrium at T_c

$(\sum S_i)/N \neq 0$. Owing to spontaneous fluctuations, the magnet has chosen one of the two possible directions, and with few exceptions the spins S_i are aligned parallel to this direction. This means that we observe a non-zero spontaneous magnetization, and even if we were to average $\sum S_i$ over a time interval t_1, $\langle (\sum S_i)/N \rangle_{t_0, t_1}$ would not vanish. Does this contradict the exact statement that a finite system cannot have any phase transitions ($M(T) = 0$ for $T > 0$)?

To understand this, we have to realize the time scale at which the system can change the direction of its magnetization. At low temperatures this time can be estimated by using Arrhenius' law. To change the sign of $\sum S_i$, an energy barrier of the order $\Delta E = JL$ has to be scaled, since a wall between positive and negative magnetization has to form, which moves through the entire lattice by diffusion. This takes a time of the order $\tau \simeq t_0 \exp(\Delta E/k_B T)$, where t_0 is the relaxation time of an individual spin. In our case ($L = 100$, $k_B T = J$, $t_0 \simeq 10^{-4}$s), therefore, we obtain $\tau \simeq 10^{40}$ s, much longer than the universe has existed. Consequently, we will not be able to observe a reversal of the magnetization at $T = J/k_B$ and we have $\langle M \rangle_{t_0, t_1} \neq 0$.

At $T = 1.5 T_c$ the spins appear to be distributed almost randomly, the correlation length ξ has nearly reached the value a. With one keystroke one can now suddenly cool down the magnet to a temperature well below T_c. It should not take too many attempts to generate regions with positive and negative magnetization at the same time (Fig. 5.11). Then the result will be diffusing domain walls which annihilate each other or contract until thermal equilibrium is reached, with just a single domain left. At the critical point, the magnetization exhibits strong fluctuations as a function of time (program ising_g). According to statistical mechanics it is exactly these fluctuations of M that determine the magnetic susceptibility χ, which at T_c is a divergent

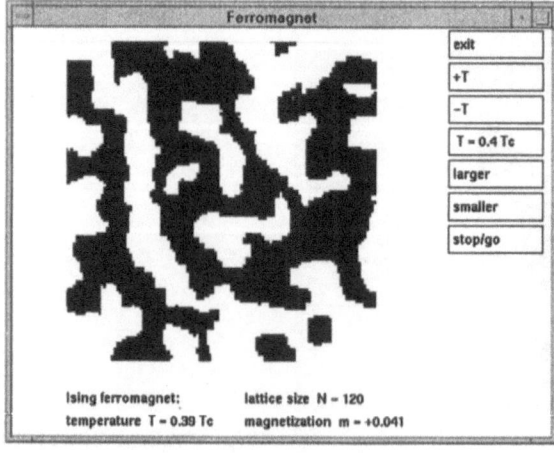

Fig. 5.11. Diffusing domain walls of the ferromagnet after sudden cooling

function of L, according to (5.58). The same is true for the energy fluctuations, which determine the specific heat C.

For $T > T_c$ and for long times t the correlations in time (just like the spatial ones) decay exponentially,

$$\langle S_i(t_0)\, S_i(t_0+t)\rangle \sim e^{-t/\tau} \; . \tag{5.64}$$

At T_c, the relaxation time $\tau(T)$ diverges as

$$\tau \sim |T - T_c|^{-z/\nu} \; , \tag{5.65}$$

where z is a new universal critical exponent whose value is only known numerically, even for the two-dimensional model ($z \simeq 2.1$). Near T_c, the fluctuations of the magnetization not only become larger, they also become slower. Therefore, this effect is called *critical slowing down*.

Exercise

The phase transition of a lattice gas is to be simulated using the Monte Carlo method. To do this, we consider a model for an $N \times N$ lattice, in which the particles can only be at the lattice sites. This is described by the variable n_i which takes the value $n_i = 1$ (0) if the lattice site i is occupied (vacant). The number of particles is controlled by a chemical potential. The dynamics consist of the ability of particles to vanish and to reappear. Neighboring particles are to repel each other so strongly that no two neighboring sites can be occupied at the same time.

Each site i is to be occupied with a probability proportional to z^{-n_i}, where z is called the *fugacity*. Other than that, the model does not contain any interactions and consequently does not have a temperature either; the only control parameter is the fugacity z, which adjusts the density of particles on the lattice.

If the lattice is maximally covered, $N^2/2$ particles are located on one of two sublattices, either on the "white" or the "black" sites, if we visualize the lattice as a checkerboard. But even when the coverage is incomplete the particles prefer one of the sublattices; in the thermal average, the difference between the number of particles on the white sites and those on the black ones is non-zero. This difference is the order parameter of the system; it corresponds to the magnetization in the Ising model. Only if the degree of coverage falls below a certain level will the two sublattices be occupied with equal density in the thermal average; in that case, the order parameter has the value zero. As in the Ising model, this phase transition only occurs in the infinite system, strictly speaking; in the simulation, one has to use finite size scaling in order to calculate the properties of the transition.

Program the following Monte Carlo step for the lattice gas:

1. Randomly select a lattice site.
2. If the lattice site is vacant, occupy it if this is permitted, i.e., if all nearest neighbor sites are vacant; if it is not permitted, accept the old configuration.
3. If the selected site is occupied, draw a random number $r \in [0, 1]$ and compare it to z^{-1}. If r is smaller, remove the particle; otherwise, accept the old configuration again.

Calculate the mean value of the order parameter and its variance, the susceptibility, as a function of the fugacity z. Determine its critical value z_c and the corresponding degree of coverage ρ_c of the lattice, where ρ is the ratio of the number of particles to the number of lattice sites.

Hint: In the literature, one finds the values $z_c = 3.7959$ and $\rho_c = 0.36776$.

Literature

Binder K., Heermann D.W. (1992) Monte Carlo Simulation in Statistical Physics: An Introduction. Springer, Berlin, Heidelberg, New York

Gould H., Tobochnik J. (1996) An Introduction to Computer Simulation Methods: Applications to Physical Systems. Addison-Wesley, Reading, MA

Honerkamp J. (1994) Stochastic Dynamical Systems: Concepts, Numerical Methods, Data Analysis. VCH, Weinheim, New York

Koonin S.E., Meredith D.C. (1990) Computational Physics, Fortran Version. Addison-Wesley, Reading, MA

5.6 The Traveling Salesman Problem

A traveling salesman wants to visit a large number of cities in as short a time as possible. So he plans an itinerary, i.e., an order in which he wants to visit these cities, that leads to as short a route as possible. What is the shortest round trip during which each city is visited exactly once? This seemingly simple problem, the *traveling salesman problem* (TSP), is the standard example of the difficult problems of combinatorial optimization. A generation of mathematicians and engineers has analyzed it mathematically and developed algorithms for its solution.

Surprisingly, the minimization of the round trip can also be considered as a problem of statistical mechanics. The total distance covered during the trip corresponds to the energy of a multiparticle problem, and with the Monte Carlo method from the previous section we can in principle find the shortest round trip, in practice a very short one. The method by which one tries to approach the ground state has been termed *simulated annealing*.

Physics

Distribute N points – each point corresponds to a city – on a square with sides of length L and mark a round trip that touches each point exactly once. How many such round trips are there? A given route is defined by the order of the points, and obviously there are $N!$ different arrangements, namely all permutations of the N points. Since each route can be started at any of the N points and traveled in either direction without changing its length, $(N-1)!/2$ different routes have to be considered. For $N = 100$, one obtains about 10^{155} routes, much more than one will ever be able to check out on the computer.

$N!$ increases faster than e^N and very much faster than any power of N. One can easily figure out a regular arrangement of cities for which the shortest round trip can be specified, but mathematicians are almost certain that in the worst case there is no algorithm that can calculate the shortest route in a number of steps that increases as a polynomial in N. The factorial $N!$ increases as $N^{1/2}(N/e)^N$. Using the methods of computer science, one can, however, reduce the task of counting out the paths to $N^2 2^N$ computation steps, which for $N = 100$ amounts to 10^{34} calculations. But this is still too many, even for one of today's supercomputers.

Although the numerical solution of the problem appears almost hopeless, computer science has developed methods with which, in one concrete case, the shortest round trip was exactly calculated for a traveling salesman who is to visit approximately $N = 2500$ cities. Mathematically, the TSP can be formulated as follows: let c_{ij} be the distance between the cities i and j, and let $x_{ij} = 1$ if the itinerary involves a direct connection between the cities i and j, $x_{ij} = 0$ otherwise. Both c_{ij} and x_{ij} are thus symmetric with respect to the indices i and j. Then we are looking for the smallest length

$$E_0 = \min_{\{x_{ij}\}} \sum_{i<j} c_{ij} x_{ij} \qquad (5.66)$$

with the constraint

$$\sum_{j=1}^{N} x_{ij} = 2 \quad \text{for all } i . \qquad (5.67)$$

Equation (5.67) means that the trip takes each city into account exactly once and that correspondingly each city has exactly two immediate neighbors. To exclude disjoint subitineraries, one also has to require

$$\sum_{i \in S} \sum_{j \notin S} x_{ij} \geq 2 \quad \text{for all } S , \qquad (5.68)$$

where S is any true subset of the indices $(1, 2, \ldots, N)$ except the empty set. The successful idea for solving the TSP extends this discrete optimization problem ($x_{ij} \in \{0, 1\}$) to a continuous one ($x_{ij} \in \mathbb{R}$), which can be solved with the standard methods of linear optimization similar to what we discussed in Sect. 1.9. By adding more and more constraints one can eventually restrict all the continuous variables x_{ij} to the allowed values 0 and 1. With this, one has solved the problem exactly.

The algorithm is very complicated and requires extensive programming. Here we want to present a different method for solving the TSP, which evolved in 1982 from the statistical mechanics of disordered magnetic systems (*spin glasses*) and is relatively easy to program. To do so, we interpret the variables $x_{ij} \in \{0, 1\}$ as the degrees of freedom of a multiparticle system with the energy $H(\underline{S})$:

$$H(\underline{S}) = \sum_{i<j} c_{ij} x_{ij} . \tag{5.69}$$

$\underline{S} = (x_{12}, x_{13}, \ldots x_{N-1,N})$ designates one of the $(N-1)!/2$ possible round trips, and H is the length of that trip. We are looking for the state with the lowest energy. Physics, or more specifically the thermodynamics of solids, tells us how we should proceed: We have to heat up the system and then cool it down very slowly. If the system is always in thermal equilibrium, then it has to reach the state of lowest energy for $T \to 0$. Also, the previous section tells us how we can accomplish the heating and cooling in this problem, namely with the help of a Monte Carlo simulation. In doing so, a stochastic process is constructed with a transition probability $W(\underline{S} \to \underline{S}')$ which obeys detailed balance with the Boltzmann factor from (5.49), $P(\underline{S}) = \exp(-H(\underline{S})/T)/Z$. T is a formal "temperature" which has to be specified in the same units as $H(\underline{S})$, i.e., as a length.

In the previous section we have seen that the sequence of round trips $\underline{S}(0), \underline{S}(1), \ldots, \underline{S}(t)$ generated this way relaxes to thermal equilibrium. This means that the probability of finding the path $\underline{S}(t)$ will be $P(\underline{S}(t))$, if only t is large enough. Now we can cool down the system so slowly that it is in thermal equilibrium all the time. Obviously $P(\underline{S})$ will only be different from zero at $T = 0$ if $H(\underline{S})$ corresponds to the lowest energy (= the shortest route).

Thus, we are certain to find the shortest round trip if only we cool down the system slowly enough. But how slow is "slow enough?" This question is difficult to answer. If we wander through the space of all routes $\underline{S}$ and regard $H(\underline{S})$ as an altitude, then $H(\underline{S})$ resembles a complex mountain range with many valleys, side valleys, gorges, and pits. So if we only move downhill, there is a high probability that we will end up in a secondary valley far from the absolute minimum. On the other hand, the Monte Carlo method (at $T > 0$) permits uphill steps as well, with a probability that depends on the difference in altitudes. But how long does it take to scale a high ridge in order to reach a deeper valley?

As in the previous section, this time can be estimated by using Arrhenius' law. Let $\Delta E > 0$ be the difference in altitude to be scaled. Then a typical time is

$$\tau = \tau_0 e^{\Delta E/T} . \tag{5.70}$$

When cooling down, we have to give the wanderer enough time to cross the high ridges. By inverting (5.70) we see that a cooling schedule $T(t)$ laid out for the simulation – here, t is the actual processing time – should fulfill the inequality

$$T(t) \geq \frac{\Delta E}{\ln(t/\tau_0)} \tag{5.71}$$

if possible. Therefore, as a function of time the temperature T may only sink at an extremely slow rate, in order to guarantee thermal equilibrium. In practice one will never be able to invest enough processing time to perfectly simulate thermal equilibrium. But if only we can stay close to the equilibrium, we will still find an energy that almost reaches the ground state energy. Consequently, the path length will be almost optimal, while the actual path may look entirely different from the optimal one.

With little programming effort and little knowledge about the problem, but with a lot of computing time, one can therefore find a good local minimum for a complex problem of combinatorial optimization with this method of simulated annealing. The method is easily applied to many other problems of this kind; at present, however, it is facing competition from genetic optimization algorithms, which work with populations of paths.

Before we discuss the details of the algorithm, we want to estimate the shortest round trip for the case of N cities distributed randomly and without correlation over a square with sides of length L. If we calculate for each city i the distance r_i to its nearest neighbor, then obviously the length of the shortest round trip is limited by the inequality

$$E_0 \geq \sum_{i=1}^{N} r_i . \tag{5.72}$$

For $N \to \infty$, the right-hand side can be written as $N\langle r \rangle$, where $\langle r \rangle$ is the average distance between nearest neighbors. We now want to calculate $\langle r \rangle$, neglecting boundary effects; to this end we consider a sufficiently large section of the plane, in which cities are distributed with a density $\rho = N/L^2$. Let $w(r)dr$ be the probability that the nearest neighbor has a distance between r and $r + dr$. In other words, $w(r)dr$ is the probability of not finding a nearest neighbor between 0 and r, multiplied by the probability of finding a city in the circular ring with the area $2\pi r dr$:

$$w(r)\, dr = \left[1 - \int_0^r w(r')\, dr' \right] \rho 2\pi r dr . \tag{5.73}$$

This integral equation for $w(r)$ can immediately be transformed into a differential equation for $f(r) = 1 - \int_0^r w(r')dr'$ by using $f'(r) = -w(r)$, which results in

$$f' = -2\pi\rho r f \quad \text{with} \quad f(0) = 1 . \tag{5.74}$$

For this we obtain the solution

$$f(r) = e^{-\pi\rho r^2} \quad \text{and} \quad w(r) = 2\pi\rho r e^{-\pi\rho r^2} . \tag{5.75}$$

The average distance between nearest neighbors can be calculated using $w(r)$:

$$\langle r \rangle = \int\limits_0^\infty r w(r)\, dr = 2\pi\rho \int\limits_0^\infty e^{-\pi\rho r^2} r^2 dr = \frac{1}{2\sqrt{\rho}} = \frac{L}{2\sqrt{N}} . \tag{5.76}$$

Then, according to (5.72), the following relation holds for the length E_0 of the shortest round trip (in the limit $N \to \infty$):

$$E_0 \geq \frac{1}{2} L\sqrt{N} . \tag{5.77}$$

If one takes into account the next-to-nearest neighbors correspondingly, then the coefficient $1/2$ changes to $0.599\dots$. In any case, though, one obtains the scaling behavior $L\sqrt{N}$, so one can compare different system sizes L and numbers of cities N. At present, the exact value of the normalized length

$$\ell = \frac{E_0}{L\sqrt{N}} \tag{5.78}$$

can only be determined numerically; for example, Percus and Martin find $\ell = 0.7120 \pm 0.0002$ in the limit $N \to \infty$.

Algorithm

As in the previous section we have to find a simple way to generate changes of the state $\underline{S}$. Allowed paths can be represented by an array that indicates the order in which the cities are visited,

$$\underline{S} = (i_1, i_2, \dots, i_N) , \tag{5.79}$$

where $i_k \in \{1, \dots, N\}$ is the number of the city. There are many possible ways of turning a round trip $\underline{S}$ into a slightly changed path $\underline{S}'$. For the sake of simplicity, we will only present the following one: we randomly select a position p and a length $l \leq N/2$, cut the segment $i_p \dots i_{p+l}$ out of the path $\underline{S}$, and reinsert it in reverse order,

$$\underline{S}' = (i_1, \dots, i_{p-1}, i_{p+l}, i_{p+l-1}, \dots, i_p, i_{p+l+1}, \dots, i_N) . \tag{5.80}$$

Here, $p + l$ is to be understood as modulo N. This move is displayed in Fig. 5.12. Using $N = 6$, $p = 3$, and $l = 2$, the round trip $\underline{S} = (1, 2, 3, 4, 5, 6)$ turns into $\underline{S}' = (1, 2, 5, 4, 3, 6)$. By applying this kind of move repeatedly, we can obviously move through the entire set of all possible paths and eliminate all crossovers. This gives us everything we need in order to use the Monte Carlo simulation described in the previous section. Any attempted move is accepted if the new round trip is shorter $\left(H(\underline{S}') \leq H(\underline{S})\right)$ or if a random number $r \in [0, 1]$ is less than the Boltzmann factor $\exp\left[-\left(H(\underline{S}') - H(\underline{S})\right)/T\right]$.

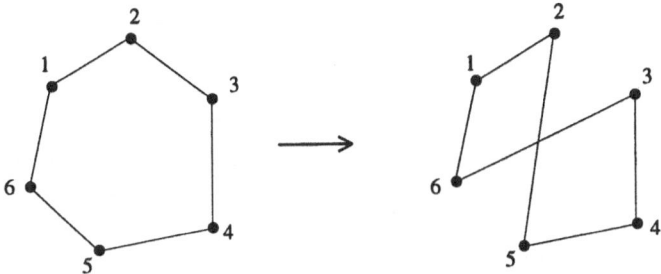

Fig. 5.12. Step-by-step change of the round trip in the Monte Carlo simulation

In the latter case, we have to calculate both a random number r and the function exp($\ldots$) each time. One can avoid both by replacing the Boltzmann factor with the step function $\theta\left[T - \left(H(\underline{S}') - H(\underline{S})\right)\right]$. In this case the move $\underline{S} \to \underline{S}'$ will always be accepted if the energy difference $H(\underline{S}') - H(\underline{S})$ is less than T. We do not know of any theoretical justification for this simplification, but lately it has been used successfully in combinatorial optimization.

The C program `travel.c` takes both methods into account. N cities are randomly distributed in a square of size L. The array map of structures `city` is used to store the coordinates (x, y) of each city:

```
typedef struct {int x,y;} city;
city map[MAXCITIES];
```

N is set to 100, but alternatively it can be passed as a parameter when calling `travel`. This is done via

```
main(int argc,char *argv[])
{
    ...
    if(argc>1) N=atoi(argv[1]);
    ...
}
```

The order in which the cities are visited is stored in the array `int path[N]`. At the beginning of the program the coordinates (x, y) of the cities are selected randomly, and the initial path is set to $\underline{S} = (0, 1, \ldots, N-1)$. Then the main loop looks as follows:

```
while (!done)
  {
    event();
    anneal(path);
    if(count++ > DRAW)
      {
        drawpath(path);
        oldlength = length(path);
        print(oldlength);
        if(aut) temp = .999*temp;
```

```
      count=0;
    }
  }
```

The function **event()** examines the keyboard buffer. While the program is running, the following actions are possible: increasing or decreasing the temperature by 10%, choosing between the methods *Boltzmann factor* or *threshold function*, switching to automatic temperature decrease, and storing the current round trip configuration. Finally, one can make the temperature jump to **temp=0**. In that case the program will only accept those moves that reduce the path length.

The function **anneal(path)** calculates the next Monte Carlo step. The position **pos** and length **len** which are needed in order to generate the new path via **change(newpath,oldpath,pos,len)** are selected randomly. Furthermore, we define a function **changed_length(...)** in order to determine the length of the new path. This is programmed most effectively by calculating just the length difference with respect to the old configuration. To prevent an accumulation of roundoff errors, the path length is recalculated from scratch at regular intervals via **length(path)**. The new path is accepted if the difference **de** of the two path lengths obeys the condition

```
(ann == 1 && (de < 0 || frand() < exp(-de/temp)))   ||
(ann == 0 && de < temp)
```

If this expression is true, i.e., has the value 1, the move $S \to S'$ is accepted and the standard function **memcpy** is used to copy the new path to the memory location of the old one. Altogether, the function is realized as follows:

```
void anneal(int oldpath[])
  {
    double newlength,de,lscal;
    int pos,len,newpath[MAXCITIES];

    pos=rand()*f1;
    len=rand()*f2;
    change(newpath,oldpath,pos,len);
    newlength=changed_length(oldlength,oldpath,pos,len);
    de=newlength-oldlength;
    if(de==0.) return;
    if((ann==1 && (de < 0 || frand() < exp(-de/temp))) ||
       (ann==0 && de < temp))
    {
      memcpy(oldpath,newpath,N*sizeof(int));
      oldlength=newlength;
      flipcount++;
    }
  }
```

f1 and **f2** are previously defined scale factors and **frand()** is a random number generator which generates real numbers distributed uniformly in the unit interval. It can be defined at the beginning of the program, for example via

```
#define frand() (double)rand()/(RAND_MAX+1.)
```

The new path is generated by first copying the old path (op) to the new one (np). Then the order of the cities between pos and pos+len is reversed. The code for this is

```
void change(int np[],int op[],int pos,int len)
{
  int i,j;

  memcpy(np,op,N*sizeof(int));
  j=len;
  for(i=0;i<=len;i++)
    {
      np[(pos+i) % N] = op[(pos+j) % N];
      j--;
    }
}
```

The algorithm for calculating the path length is obvious:

```
double length(int path[] )
{
  int i,j;
  double l=0.,dx,dy;

  for(i=0;i<N;i++)
    {
      j=(i+1)%N;
      dx=map[path[i]].x-map[path[j]].x;
      dy=map[path[i]].y-map[path[j]].y;
      l += sqrt(dx*dx+dy*dy);
    }
  return (l);
}
```

Additional functions called drawpath and print draw the path and write text to the graphics window respectively. After compiling the program we can use the command travel to watch the search on the screen and increase or decrease the temperature in the process.

Results

We have read the data from travel.dat into *Mathematica* and plotted them there. Figure 5.13 shows the randomly generated initial state of about 100 cities. The normalized path length ℓ from (5.78) has the value $\ell \simeq 4.8$. We start with a temperature L/8. Even at this relatively high temperature the system quickly relaxes to the value $\ell \simeq 2.2$, with about 20% of the attempted

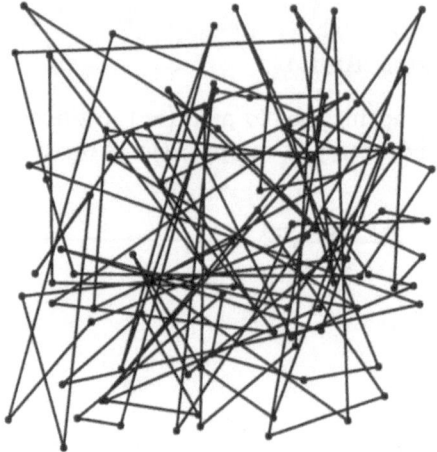

Fig. 5.13. Randomly generated initial state with a normalized path length $\ell = 4.8$

moves being accepted. A slow automatic cooling process in which the temperature is reduced by 0.1% for every 100 steps brings the system to the states shown in Figs. 5.14 and 5.15 respectively after a processing time of 30 minutes on our PC. Both states are local minima of the energy landscape, i.e., any changes $\underline{S} \to \underline{S}'$ will always lead to longer paths, which are no longer accepted at $T = 0$. In this case – and in most others – the fast threshold method yielded a slightly lower value ($\ell = 0.858$) than the slower Metropolis algorithm ($\ell = 0.883$). If one only moves downhill in the complex energy landscape $H(\underline{S})$ right from the beginning (temp=0), then the path will be longer ($\ell = 0.909$), but one will obtain that result much faster than with the two cooling methods. Figure 5.16 compares the results from Figs. 5.14 and 5.15. Although the lengths of the two round trips are almost the same, the itineraries are clearly different from each other. If one repeats the simulation with different random numbers (for the same cities), then one obtains almost the same path length, but again a completely different order of the cities.

We have verified the normalization (5.77) by simulating a round trip with 300 cities and slowly cooling down with the threshold algorithm. The result, $\ell = 0.826$, confirms the normalization $E_0 \propto \sqrt{N}$.

Of course, we will never know whether we have really found the shortest path. To answer this question we need the relatively involved methods of computer science. But one can certainly recommend simulated annealing as a method that requires little effort to obtain a good solution for an optimization problem about which much is yet unknown.

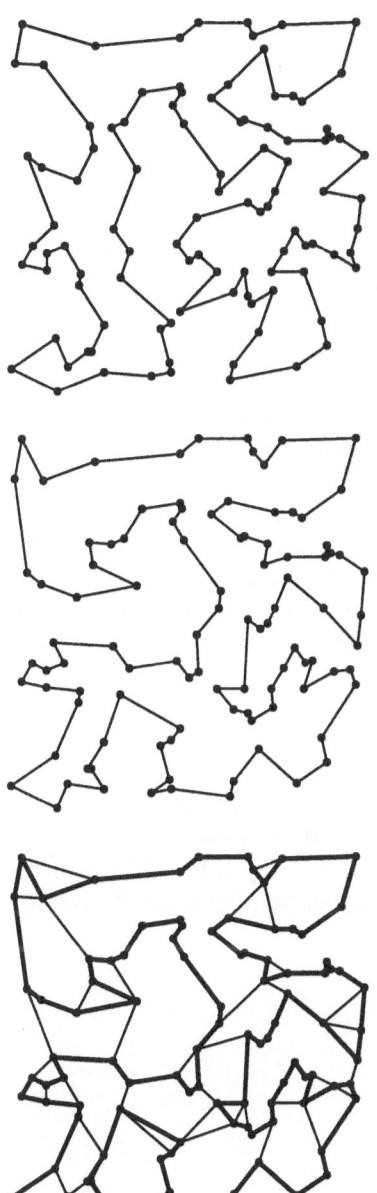

Fig. 5.14. The round trip after simulated annealing with the Boltzmann algorithm. Now the normalized path length is $\ell = 0.883$

Fig. 5.15. The round trip for the same problem as in Fig. 5.14, but generated with the threshold algorithm. The result is a slightly shorter path with $\ell = 0.858$

Fig. 5.16. Comparison of the two optimized paths from Figs. 5.14 (*thin lines*) and 5.15 (*thick lines*). Though the two curves have almost the same length, they differ significantly from each other

Exercise

Here the reader is to try and solve the optimization problem from Sect. 1.3 with the methods of simulated annealing. A signal with a given power spectrum is to be transmitted, keeping the peak voltage as low as possible.

The voltage is probed at $N = 64$ discrete points in time, resulting in a sequence $U_1, \ldots, U_N$ whose Fourier transform $\{b_1, \ldots, b_N\}$ or more precisely whose power spectrum is known: $|b_s|^2 = |b_{N-s+2}|^2 = 1$ for $s = 9, 10, \ldots, 17$, $b_s = 0$ otherwise. To ensure that the voltage signal $\{U_1, \ldots, U_N\}$ is real, we have to require $b_{N-s+2} = b_s^*$.

We are looking for the phases φ_s which, with $b_s = e^{i\varphi_s}$, minimize the peak voltage; in other words, the value of $\min_{\{\varphi_s\}}(\max_r |U_r|)$ is to be determined.

Now the state $\underline{S}$ from Sect. 5.5 is a configuration of possible phases $(\varphi_9, \ldots, \varphi_{17})$, and the quantity which corresponds to the energy is the peak voltage $H(\underline{S}) = \max_r |U_r|$. In each Monte Carlo step the configuration of phases is modified randomly, for example by varying one randomly selected phase. This results in an energy change, and the algorithm from Sect. 5.5 then decides whether the new configuration is accepted.

The best result so far, which was obtained by one of our students, is $H = 0.739545$. Whoever improves on this value is requested to let us know about it.

Literature

Percus A.G., Martin O.C. (1996) Dimensional Dependence in the Euclidean Travelling Salesman Problem. Phys. Rev. Lett. 76:1188

Press W.H., Teukolsky S.A., Vetterling W.T., Flannery B.P. (1992) Numerical Recipes in C: The Art of Scientific Computing. Cambridge University Press, Cambridge, New York

A. First Steps with *Mathematica*

Mathematica is started with the command **math** on almost all computers. On systems with a graphical user interface this is sometimes accomplished by clicking on an icon. *Mathematica* comes up with the prompt *In[1]:=* and awaits an input which is terminated by RETURN. The notebook version of *Mathematica* is just slightly different in that commands are sent individually or in groups with SHIFT-RETURN. *Mathematica* is an interpreter language that processes input immediately and presents the result on the screen in the form *Out[..]=* .... In doing this, all expressions which were defined earlier in the same session are taken into consideration. One ends a *Mathematica* session with *In[..]:=***Quit**. The command **Exit** or CONTROL D on Unix systems accomplishes the same thing.

The command **math** starts a *Mathematica* session.

*In[..]:=***Quit** ends the session.

Simple Arithmetic

In[1]:= **4 + 7**

Out[1]= 11

In[2]:= **3 4.2**   *space = multiplication*

Out[2]= 12.6

In[3]:= **2*3*4**   ** = multiplication too*

Out[3]= 24

In[4]:= **24/8**   */ = division*

Out[4]= 3

In[5]:= **2^3**   *^ = power*

Out[5]= 8

In[6]:= **(3+4)^2-2(3+1)**

Out[6]= 41

In[7]:= **%/5**   *% = the most recent result displayed*

$Out[7]= \dfrac{41}{5}$

$In[8]:= \mathbf{N[\%]}$ $N[x]$ = *numerical value of x*

$Out[8]= 8.2$

$In[9]:= \mathbf{22.4/(6.023\ 10^{\wedge}23)}$

$Out[9]= 3.71908\ 10^{-23}$

$In[10]:= \mathbf{3^{\wedge}-2*36+4}$ *Order of precedence: power before*
                                            multiplication before addition

$Out[10]= 8$

The Most Important Functions

$In[11]:= \mathbf{Sqrt[16]}$ $Sqrt[x] = \sqrt{x}$

$Out[11]= 4$

$In[12]:= \mathbf{N[Pi/2]}$ Pi = *Mathematica's form of* π

$Out[12]= 1.5708$

$In[13]:= \mathbf{Sin[\%]}$ $Sin[x] = \sin x$

$Out[13]= 1.$ $1. = 1.0$

All functions, procedures, constants, ...
available in *Mathematica* begin with a capital letter.
Function arguments are always placed in square brackets [].

$In[14]:= \mathbf{Cos[Pi/4]}$ $Cos[x] = \cos x$

$Out[14]= \dfrac{1}{Sqrt[2]}$ *exact value*

$In[15]:= \mathbf{Exp[0.5]}$ Exp = *exponential function*

$Out[15]= 1.64872$ *numerical value*

$In[16]:= \mathbf{Log[\%]}$ $Log[x] = \ln x$

$Out[16]= 0.5$

$In[17]:= \mathbf{?Log}$
Log[z] gives the natural Logarithm of z (logarithm to
base E). Log[b, z] gives the logarithm to base b.

$In[17]:= \mathbf{Log[E, \%15]}$ *%15 is the same as* Out[15]. *All* In[n]
                                            and Out[n] *generated during a session*
                                            can be referred to again.

$Out[17]= 0.5$

In *Mathematica* one obtains help by typing a question mark (**?**)
In[..]:= **?Log** ⟶ information about Log[x].
In[..]:= **??Log** ⟶ more information about Log[x].
In[..]:= **?L*** ⟶ all functions that begin with L.
In[..]:= **?*** ⟶ "everything" is listed.

In[18]:= **Sqrt[-1]**

Out[18]= I

In[19]:= **?I**
I represents the imaginary unit Sqrt[-1].

In[19]:= **Exp[I*N[Pi/4]]**

Out[19]= 0.707107 + 0.707107 I

In[20]:= **Re[%]+I*Im[%]**   Re[z] = *real part of z*
                          Im[z] = *imaginary part of z*

Out[20]= 0.707107 + 0.707107 I

In[21]:= **?Arc***
ArcCos ArcCot ArcCsc ArcSec ArcSech ArcSin ArcSinh
ArcTan ArcTanh ArcCosh ArcCoth ArcCsch

In[21]:= **ArcTan[1]**   ArcTan[x] = arctan x

$$Out[21] = \frac{Pi}{4}$$

Graphics

In the following we will omit the input and output designators In[..] and Out[..] at the beginning of each line. Inputs to the computer are distinguished by a boldface font, everything else is output from the computer or our commentary.

Plot[Sin[x], {x, 0, 2Pi}]   *Plot of the function* sin x.

Plot[{BesselJ[0,x], BesselJ[1,x]}, {x, 0, 10}]   *The Bessel functions* $J_0(x)$ *and* $J_1(x)$ *in one plot.*

Plot3D[Exp[-(x^2+y^2)], {x, -2, 2}, {y, -2, 2}]   *Three-dimensional plot of the two-dimensional Gaussian function* $\exp[-(x^2 + y^2)]$.

ParametricPlot[{Cos[t], Sin[t]}, {t, 0, 2Pi}]   *Parametric plot of a circle. An ellipse appears on the screen because the axis ratio is not equal to one.*

Show[%, AspectRatio -> Automatic]   *Now we get a circle.*
Show[..] *displays graphics objects with the selected options.*

ParametricPlot3D[{Cos[t],Sin[t],t/6}, {t,0,6Pi}]
A helix in three dimensions.

ParametricPlot3D[{r Cos[t],r Sin[t],r^2},
                                        {r,0,2},{t,0,2Pi}]
A rotational paraboloid as a surface in three dimensions.

ContourPlot[Exp[-(x-1)^2-y^2]+ Exp[-(x+1)^2-y^2],
                 {x,-3,3},{y,-3,3}, PlotPoints -> 60]
Contour lines. PlotPoints -> 60 *is an optional input.*

Options[Plot]   *A list of the options available with* Plot[..].

The most important plot commands:
Plot[f, {x, xmin, xmax}]
Plot3D[f, {x, xmin, xmax}, {y, ymin, ymax}]
ParametricPlot[{fx, fy}, {t, tmin, tmax}]
ParametricPlot3D[{fx,fy,fz}, {t, tmin, tmax}]
ParametricPlot3D[{fx,fy,fz}, {s,s1,s2}, {t,t1,t2}]
ContourPlot[f, {x, xmin, xmax}, {y, ymin, ymax}]

Symbols and Self-defined Functions

n = 20! $\longrightarrow$ 2432902008176640000, k! = Factorial[k] = *factorial of k.*

N[n] $\longrightarrow$ 2.4329 10^{18}, *the variable* n *has the value* 20!

a = Random[] $\longrightarrow$ 0.296851, Random[] *produces random numbers between 0 and 1. The number* 0.296851 *was assigned to the variable* a

N[a, 13] $\longrightarrow$ 0.2968512129099, *the number* a *to 13 digits. If not otherwise specified, decimal numbers are processed internally to 16 digits.*

Clear[a,n]   *Clears the assignment of the variables* a *and* n.

c = (a+b)^3 $\longrightarrow$ $(a + b)^3$

c = Expand[c] $\longrightarrow$ $a^3 + 3\,a^2\,b + 3\,a\,b^2 + b^3$

f[x_]:= Exp[-0.25 x^2] Cos[4.5 x]   *This is how functions are defined.* x_ *is a "place holder" with the name* x. *The underscore character in* x_ *is important. It may only be used on the left side of the equation. Instead of the symbol* := *we could have used the symbol* =.

f[1.5] $\longrightarrow$ 0.50882, *value of* f *for* $x = 1.5$.

`Plot[f[x],{x,-Pi,Pi}]`   *Plot of the function f defined above. Instead of x one can use any other variable name here as well. One obtains the same result with the command* `Plot[f[t],{t,-Pi,Pi}]` .

`fermi[e_,b_]:= 1/(Exp[b(e-1)]+1)`   *A function of two variables.*

`Plot[{fermi[e,10], fermi[e,Infinity]}, {e,0,1.7}]`   *Plot of two Fermi functions.* `Infinity` $= +\infty$.

The difference between `:=` and `=` explained by an example:
`a = Random[]`
The right side is evaluated and the resulting random number is assigned to the variable **a**. Every reference to **a** will yield this number.

`r:= Random[]`
The right side is not evaluated at first, only after the variable **r** is referred to again. The right side is reevaluated each time **r** is referred to. Each time a new random number is produced.

Lists

`numlist = {2,3,4}`   *A list with three numbers. Lists are written in curly braces* `{ }` .

`Head[numlist]` $\longrightarrow$ `List`. *Every object has a header.* `numlist` *is a list.*

`numlist^2` $\longrightarrow$ `{4,9,16}`. *Almost all functions are* `Listable`.

`Log[numlist] // N` $\longrightarrow$ *List of the logarithms of the elements of* `numlist`.

`Table[2, {7}]` $\longrightarrow$ `{2,2,2,2,2,2,2}`. `Table[..]` *creates lists.*

`Table[i^3, {i,5}]` $\longrightarrow$ `{1, 8, 27, 64, 125}`

`Table[1/j, {j,3,8}]` $\longrightarrow$ $\{\frac{1}{3}, \frac{1}{4}, \frac{1}{5}, \frac{1}{6}, \frac{1}{7}, \frac{1}{8}\}$

`Table[x, {x,0,2,0.25}]` $\longrightarrow$ `{0, 0.25, 0.5, ..., 2.}`

`Table[expr, {..}]` generates lists. The braces, with the general form `{i, imin, imax, di}`, are called 'iterator'. Abbreviations of the general form are possible.

`list = Table[{x, Gamma[x]}, {x,2,4,0.05}];`

> A semicolon (;) after an expression suppresses the next Out[..]. In spite of this the variables have the assigned values.

ListPlot[list]   ListPlot[..] *plots (lists of) data.*

letters = Table[FromCharacterCode[j], {j,122,97,-1}]
$\longrightarrow$ {z,y,x,w, ...}

Sort[letters] $\longrightarrow$ {a,b,c,d, ...} . Sort[..] *sorts not only letters but also numbers.*

letters[[3]] $\longrightarrow$ x , *the third element of the list* letters . *Parts of lists or of expressions are designated by double square brackets* [[..]] .

Length[list] $\longrightarrow$ 41 , *the length of the list* list .

list[[20]] $\longrightarrow$ {2.95, 1.91077}

list[[20,1]] $\longrightarrow$ 2.95

list[[20,2]] $\longrightarrow$ 1.91077

abc = Reverse[letters] $\longrightarrow$ {a,b,c,d,...}

abc[[{2,4,6}]] $\longrightarrow$ {b,d,f}

abc[[-1]] $\longrightarrow$ z . *A negative index corresponds to reverse numbering.*

Permutations[{1,2,3}] $\longrightarrow$ *all permutations of* {1,2,3} .

Storing, Reading, and Fitting of Data

Clear["Global`*"]   *Clears all self-defined variables and functions.*

f[x_] = 1.0 x^4 - 2.0 x^2 + 0.5 *A fourth-order polynomial.*

?f $\longrightarrow$ *the definition of* f[x] .

p1 = Plot[f[x], {x,-1.5,1.5}]

r:= Random[Real, {-0.1,0.1}]   Random[] *(without argument) returns a real random number between* 0 *and* 1. Random[Type,Range] *is the more general form.*

data = Table[{x, f[x]+r}, {x,-1.5,1.5,0.05}] // Chop
Noisy data. Chop *replaces real numbers which deviate from* 0 *by less than* 10^{-10} *with* 0.

p2 = ListPlot[data]

Show[p1,p2]   Show[..] *can combine graphics objects.*

`g[x_] = Fit[data, {1,x,x^2,x^3,x^4}, x]` Fit[..] *yields a linear least-square fit.*

`Plot[{f[x],g[x]},{x,-1.5,1.5},`
`         PlotStyle -> {RGBColor[0,0,0],RGBColor[1,0,0]}]`
Original curve black, fit curve red.

`data >> list.dat`   *Objects in Mathematica can be written to a file with* >>filename.

`!ls`   *This is the format for sending commands to the operating system during a Mathematica session.*

`!command`   sends the command to the operating system.
`!!filename`   displays the contents of the file `filename` on the screen.

`!!list.dat`

`data2 = << list.dat`   *The command* << *is the counterpart of* >> . *With this command one can read Mathematica objects from files.*

`data2 == data` $\longrightarrow$ True.   == *is the logical equality operator.*

`OutputForm[TableForm[data]] >> measured.dat`

`!!measured.dat`   measured.dat *could be a data file that was generated in an experiment.*

`data = ReadList["measured.dat",{Real,Real}]`   ReadList[..] *turns external data into a list, here a list of pairs of numbers.*

`model[x_] = a + b x + c x^2 + d x^3 + e x^4`   *We define a model function whose parameters we want to fit.*

`Needs["Statistics`NonlinearFit`"]`   *This command loads the* package NonlinearFit.m .

`h[x_] = NonlinearFit[data,model[x],x,{a,b,c,d,e}]` *The procedure* NonlinearFit[..] *replaces the parameters in the model function by their optimal values. The related procedure* NonlinearRegress *yields the best fit parameters themselves and more statistical information.*

Simple *Mathematica* Programs

We assume that a file `program1.m` with the following contents has been generated using a text editor:

```
(* This is the format for comments in Mathematica *)

Clear[" Global`*"]                (* clears old definitions *)
f[x_] := Tan[x/4]-1
Plot[f[x],{x,2,4}]
r = FindRoot[f[x]==0,{x,3.1}]     (* 3.1 = initial value *)
pi = x /. r
```

<< **program1.m** $\longrightarrow$ *Plots* f[x] *and returns the value* 3.14159, *an approximation of* π. FindRoot[..] *finds numerical solutions of equations.*

N[pi-Pi] $\longrightarrow$ 8.88178 10^{-16}

The next program, stored as program2.m, uses the *Mathematica* function Block[..] and calculates the first k zeros of the Bessel function $J_n(x)$. It uses the fact that, for $n > 0$, the first zero of $J_n(x)$ lies approximately at $n + 1.9 \, n^{1/3}$ and that the following zeros are all separated by about π.

```
besselzero[n_,k_] :=
    Block[{f,start,rule,zero,list},
          f[x_] = BesselJ[n,x];
          start = If[n==0, 2.5, n + 1.9 n^(1/3)];
          rule = FindRoot[f[x]==0,{x,start}];
          zero = x /. rule; list = {zero};
          While[Length[list] < k,
                start = list[[-1]] + Pi;
                rule = FindRoot[f[x]==0,{x,start}];
                zero = x /. rule;
                AppendTo[list,zero] ];
          If[n > 0, PrependTo[list,0]];
          Plot[f[x],{x,0,list[[-1]]+1}];
          list ]
```

<< **program2.m**

besselzero[3,4] $\longrightarrow$ *Plot of* $J_3(x)$ *and a list of the first four zeros of* $J_3(x)$.

Loops and *If* statements

Do[Print[" Hello World"],{20}] $\longrightarrow$ *Writes* Hello World *to the screen* 20 *times.*

```
Loops:
Do[expr, {iterator}]
For[start, test, incr, body]
While[test, body]
Nest[f, expr, n]
NestList[f, expr, n]
FixedPoint[f, expr]
```

`s1 = "    " ; s2 = "   ASCII code ";`   *s1 and s2 are strings.*

`Do[Print[s1,FromCharacterCode[j],s2,j],{j,65,75}]`
Writes the first 11 capital letters and their ASCII codes to the screen.

`For[i=1, i < 11, i++, Print[i,"    " ,i^2]]` $\longrightarrow$ *Generates the numbers from 1 to 10 and their squares.*

`test:= (a = Random[]; a < 0.8)`

`While[test, Print[a]]` $\longrightarrow$ *On average four random numbers between* 0 *and* 0.8 .

`Nest[Sin, t, 3]` $\longrightarrow$ `Sin[Sin[Sin[t]]]`

`Nest[Sin, 1.5, 30]`   *As above, but iterated* 30 *times and with the initial value* 1.5 .

`NestList[Sin, 1.5, 30]`   `NestList[..]` *yields a list of all values calculated in the course of the iteration.*

`f[x_] := N[Cos[x],18]`   *numerical value of* $\cos x$ *to* 18 *digits.*

`FixedPoint[f,1/10]` $\longrightarrow$ 0.739085133215160642, *the fixed point of* $\cos x$, *i.e., the* x *for which* $\cos x = x$ *to sufficient accuracy.*

`If[Random[] < 0.5, Print["number < 0.5"]]`

`beep:= Print["\007"]; beepbeep:=(beep;Pause[1];beep)`

`If[Random[] < .5, beep, beepbeep]`

`theta[x_]:= If[x <= 0, 0, 1]`   *Definition of the* Θ *function.*

`Plot[theta[x],{x,-2,2},AspectRatio -> Automatic]`

Differentiation, Integration, Taylor Series

`Clear["Gobal`*"]`

`D[Exp[x^2], x]` $\longrightarrow$ *the derivative of* $\exp(x^2)$ *with respect to* x.

`Integrate[1/(x^4-1), x]` $\longrightarrow$ *the indefinite integral of* $1/(x^4 - 1)$.

`Simplify[D[%,x]]` $\longrightarrow$ *the integrand of the above integral. The function* Simplify[..] *reduces fractions to their least common denominator and tries to put the result into the simplest possible form.*

`Integrate[Exp[c*x],{x, a, b}]` $\longrightarrow$ *the definite integral of* e^{cx} *over x in the limits a to b.*

`Integrate[Sin[Sin[x]],{x, 0, 1}]`   *No success.*

`NIntegrate[Sin[Sin[x]],{x, 0, 1}]` $\longrightarrow$ 0.430606, *numerical integration with* NIntegrate[..].

`Integrate[x*y*Exp[-(x^2 + y^2)],{x,0,2.0},{y,0,x}]`   *A double integral whose value is* 0.120463. *It is first integrated over y. The limits of this y integration may contain the inner integration variable x.*

`series = Series[Tan[x], {x, 0, 10}]` $\longrightarrow$ *the series expansion of* $\tan x$ *about* $x = 0$ *up to the term of the order* x^{10} *plus* O[x]^{11}, *which represents the neglected higher-order terms.*

`Head[series]` $\longrightarrow$ SeriesData, *a Mathematica object in its own right, not a polynomial.*

`Normal[series]`   Normal[..] *turns this into a polynomial.*

`Timing[Series[Exp[Sin[x]], {x,0,80}];]` $\longrightarrow$ {11.54 Second, Null}

`Timing[Series[Exp[Sin[1.0*x]],{x,0,80}];]` $\longrightarrow$ {0.47 Second, Null}, *numerical results can often be calculated much faster than exact results.*

Vectors, Matrices, Eigenvalues

`Remove["Global`*"]`   Remove[..] *is even more radical than* Clear[..].

`v = {x,y,z}`   *Vectors are written as lists.*

`r = Table[x[j],{j,3}]` $\longrightarrow$ {x[1], x[2], x[3]}

`b = Table[1/(i+j),{i,3},{j,3}]`   *Matrices are represented as lists of lists.*

`MatrixForm[b]` $\longrightarrow$ *the matrix b .*

`d = DiagonalMatrix[{d1,d2,d3}]` $\longrightarrow$ *a diagonal matrix with the diagonal elements* d1, d2, d3.

`IdentityMatrix[3]` $\longrightarrow$ *the* 3×3 *unit matrix.*

Transpose[b] == b $\longrightarrow$ True *because* b *is symmetric.* Transpose[b] *yields the transpose of* b.

Det[b] $\longrightarrow$ *the determinant of* b.

v.r $\longrightarrow$ *the scalar product of* v *and* r.

b.v $\longrightarrow$ *the matrix* b *applied to the vector* v.

b.d $\longrightarrow$ *the matrix product* b *times* d.

b.Inverse[b] $\longrightarrow$ *the unit matrix;* Inverse[b] *yields the inverse of the matrix* b.

Eigenvalues[b] $\longrightarrow$ *the eigenvalues of the matrix* b. *They are the zeros of a third-order polynomial, which Mathematica can calculate with the Cardano formula.*

nb = N[b] $\longrightarrow$ *the numerical approximation of the matrix* b. *The exact rational numbers are replaced by decimal numbers with an accuracy of* 10^{-16}.

Eigenvalues[nb] $\longrightarrow$ {0.875115, 0.0409049, 0.000646659}, *a list of the eigenvalues of the matrix* nb.

u = Eigenvectors[nb] $\longrightarrow$ *a list of the eigenvectors of* nb.

Chop[u.Transpose[u]] $\longrightarrow$ *the unit matrix, since the eigenvectors of* nb *are orthonormal, except for numerical inaccuracies of the order* 10^{-16}.

Sort[Thread[Eigensystem[nb]]] $\longrightarrow$ *eigenvalues and associated eigenvectors, sorted according to the magnitude of the eigenvalues.*

Solving Equations

FindRoot[Cos[x]==x,{x,1}] $\longrightarrow$ {x -> 0.739085}, *the solution of the equation* $\cos x = x$. *The brackets* {x,1} *mean: seek a solution with respect to x and start the search at x=1.*

FindRoot[{Cos[a*x]==x,-a*Sin[a*x]==-1}, {x,1}, {a,1}]
Systems of equations can also be solved with FindRoot[..].

Plot[Zeta[x]-2,{x,1.2,3}] $\longrightarrow$ *a smooth function which has a zero in the range under consideration.*

FindRoot[Zeta[x]-2==0,{x,2}] $\longrightarrow$ *an error message because the function cannot be differentiated symbolically.*

FindRoot[Zeta[x]-2==0,{x,{1.2,3}}]
This works. If one specifies a start interval instead of an initial value, a variant of the secant method is used.

FindRoot[lhs == rhs, {x, x0}] finds numerical solutions.

Solve[eqns, vars] tries to find exact solutions.

NSolve[eqns, vars] numerical solutions of algebraic equations.

LinearSolve[m, b] solves the system of equations m.x == b

DSolve[eqn, y[x], x] solves differential equations.

NDSolve[eqns, y, {x, xmin, xmax}] finds numerical solutions of differential equations.

Solve[a*x^2+b*x+c==0,x] $\longrightarrow$ *the 2 solutions of the quadratic equation.*

Solve[ArcSin[x]==a,x] $\longrightarrow$ {{x -> Sin[a]}}

Solve[x^5-5x^2+1 == 0, x] *Too difficult.*

NSolve[x^5-5x^2+1 == 0, x] $\longrightarrow$ *5 numerical values for the zeros of this equation.*

DSolve[y''[x]==y[x], y[x], x] $\longrightarrow$ *the general solution of this differential equation.*

DSolve[y'[x]==Cos[x*y[x]], y[x], x] *No success.*

NDSolve[{y'[x]==Cos[x*y[x]],y[0]==0},y[x],{x,-5,5}] $\longrightarrow$ *a Mathematica object called* InterpolatingFunction *.*

f[x_] = y[x] /. % *This is the format that is used to convert the object* InterpolatingFunction *to a normal function which can be plotted with* Plot[f[x],{x,-5,5}] *.*

Patterns, Ordering, Sorting

Remove["Global`*"]

list = {1, 2, f[a], g[b], x^n, f[b], 3.4, Sin[p]^2}

Position[list, f[_]] $\longrightarrow$ {{3},{6}}, *a list of all positions with* f[*anything*].

Cases[list,_^2] $\longrightarrow$ {Sin[p]^2}, *all cases with the exponent 2.*

Count[list,_^_] $\longrightarrow$ 2, *the number of all cases with any exponent.*

DeleteCases[list,_[b]] $\longrightarrow$ {1,2,f[a],x^n,3.4,Sin[p]^2}, *any list items ending in* [b] *are eliminated.*

`Select[list,IntegerQ]` $\longrightarrow$ $\{1,2\}$, *the cases for which* `IntegerQ[..]` *yields the value* True. *There are more than* 30 *functions of the form* `*Q` *with which one can pose questions.*

`?*Q` $\longrightarrow$ *all these functions.*

`Select[list,NumberQ]` $\longrightarrow$ $\{1,2,3.4\}$. `NumberQ[`*number*`]` *yields the logical value* True .

`Select[list,!NumberQ[#]&]` $\longrightarrow$ *everything except the numbers. The exclamation point* (!) *is the logical negation. The construction* expr`[#]&` *turns* expr *into an operator, a so-called* pure function. # *is replaced by the argument.*

`term = Expand[3*(1+x)^3*(1-y-x)^2]` Expand`[..]` *multiplies out.*

`Factor[term]` $\longrightarrow$ *the argument of the function* Expand`[]` *above. In some ways,* Factor *is the counterpart of* Expand.

`FactorTerms[term]` *Factors out numerical coefficients.*

`Collect[term,y]` *Sorts according to powers of* y.

`Coefficient[term,x^2*y^2]` $\longrightarrow$ 2, *the coefficient of* $x^2 y^2$.

`CoefficientList[term,x]` $\rightarrow$ *the list of the coefficients of the powers of* x .

`Apart[1/(x^4-1)]` $\longrightarrow$ *the partial fraction decomposition of* $1/(x^4 - 1)$.

`Together[%]` Together`[..]` *reduces to the least common denominator.*

`N[Sort[{Sqrt[2],2,E,Pi,EulerGamma}]]` $\longrightarrow$
$\{2., 1.41421, 2.71828, 0.577216, 3.14159\}$. *According to what criteria is this sorted?*

`smaller:= (N[#1] < N[#2])&` *sorts according to the numerical value.*

`N[Sort[{Sqrt[2],2,E,Pi,EulerGamma},smaller]]` *Now it is correct.*

B. First Steps with C

C is the programming language of the Unix computers. While Fortran is mainly used for extensive numerical calculations, C allows for a clear, compact, and function-oriented programming style and the use of machine-oriented graphics interfaces. In addition, C (as does Pascal) has the advantage that there is a convenient environment for program development with a built-in graphics library for the PC.

C also has disadvantages, however: for one, it was not developed for scientific calculations. Powers are calculated with a function, the mathematics library has to be loaded each time when compiling, and complex numbers and vector and matrix multiplications have to be defined by the user. Secondly, one often has to use memory addresses, and C does not control whether indices accidentally exceed the previously defined range. Beginners in particular have to be very careful about the last point: C tolerates many mistakes and it takes a large amount of work to find the reason for obviously false results. On the other hand, C gives the programmer a lot of freedom which can be used to one's advantage.

We want to give an introduction to C, mainly with simple examples. A program must always contain the function **main( )**, whose instructions are in the subsequent brackets {...}. The types of all variables and functions have to be declared, for example with

```
int i, answer;
char ch;
float f, rational_number;
long long_i;
double long_f;
```

C distinguishes between capital and lower-case letters. The variables following int, char, and long are integers; those following **float** and **double** are rational numbers. The different types correspond to different accuracies and different sizes of the memory space assigned; the actual value of either depends on the computer. Normally one byte (= 8 bits = memory needed for an ASCII character) is reserved for char, two bytes for int, and four bytes for long.

For the first example we want to print the numbers 4, 5, ..., 15 on the screen: in a file print.c we enter the lines

```
main( )
  {
    int i;
    for( i=4; i<16; i=i+1) printf("i=%d \n",i);
  }
```

Every command has to be terminated by a semicolon. The iteration loop **for**
contains the following structure in parentheses: (initialization; termination
condition; increment or, more precisely, command to be executed after each
pass). The parentheses are followed by the command to be executed; several such commands must be grouped by braces {...} and there must be a
semicolon after every command.

This program is compiled with

```
cc  -o print print.c
```

and the command **print** writes the numbers on the screen. The print function
printf expects as its first argument a character string "..." which contains
the text as well as the formatting instructions for the variable i to be printed.
Depending on the types of the variables the following formatting instructions
are commonly used.

Format   Variable type

%d	int	
%ld	long	
%c	char	
%f	float	(floating-point representation)
%e	float	(exponential representation)
%lf	double	
%s	string	

The symbol \n is a control character that causes a line feed. It corresponds
to the RETURN key.

The formatting instructions above can also be used for reading, which is
accomplished by the function **scanf( )**:

```
main( )
  {     double i = 0.;
        for(; i<100.;)
          {   printf( "Please enter a number \n");
              scanf( "%lf", &i);
              printf( "You have entered %lf",i);
          }
  }
```

This program demonstrates some peculiarities: 1. In the type declaration the
variable can be initialized at the same time. 2. Empty initialize and increment
instructions can be used in the **for** loop. 3. If one passes a variable, here i,

to a function, in this case to scanf, only its value is passed. Consequently, scanf cannot write a new value to the memory location of i. This can only be done if the function receives the address (not the value) of the variable; then scanf can write the number entered to the memory location of i. One refers to addresses by placing a & in front of the variable name.

The program continues reading numbers and writing them to the screen until a number larger than 100 is entered. If scanf cannot convert the number entered to a real number, then the function waits for the next input.

Of course one can also read and print text. To do this, one has to reserve one memory location of type char for every character, e.g., in the form of a vector.

```
main()
{   char str[100];
    printf( "What is your name? ");
    scanf( "%s", str);
    printf( "Hello %s", str);
}
```

str[100] is a vector with room for 100 characters. str[0] is the first and str[99] the last character. str is the address of the vector (= the address of the first element, or &str[0]); therefore it is sufficient to pass the name of the vector without & to scanf. The function reads all characters until it encounters a space or a RETURN; then the end of the character string is marked with the ASCII character \0 (= null). printf prints all characters up to the character \0.

After these two first examples, we want to provide a list of commands, followed by explanatory remarks, with which one's first, small programs should be feasible.

```
#include <math.h>
#include "myfile"
#define  N 1000
```

These instructions are executed before the compilation. #include includes the specified files at the position in the code where the directive is found; <...> searches special paths, "..." searches the current path where the program is located. All functions used must be declared; that is done in the *header* files math.h, stdlib.h, graphics.h, time.h, etc. The line #define N 1000 causes the symbol N to be replaced by the symbol 1000 throughout the program.

```
sum = a+b;
a = b = a/b * e;
mod = a % b;
```

The equals sign assigns the value on the right to the variable on the left. % is the modulo operation, i.e., mod is assigned the remainder of a/b.

```
sum += a;
a++; b--;
```

C contains abbreviations for some of the most common arithmetic operations: the commands above replace the expressions sum = sum+a;, a = a+1;, and b = b-1; respectively.

```
a>b;            greater than
a>=b;           greater than or equal to
a==b;           equal to
a!=b;           not equal to
a<=b;           less than or equal to
a<b;            less than
```

If any of the above relations is true, the corresponding expression is assigned the value 1, otherwise it is assigned the value 0. There is no special data type for logical expressions; C simply works with integers of type int.

```
a<b || c!=a     or
a<b && c!=a     and
!(a<b)          not
```

Logical operators combine logical values and produce 1 (true) or 0 (false). The logical negation ! turns a 1 into a 0 and vice versa.

```
if(b==0)    printf ("Division is not defined");
   else     printf ("Result = %lf", a/b);
if (b>1.e-06)
      { scanf("%lf",&a);
        printf("a divided by b=%lf", a/b);
      }
```

The instruction if executes the following command only if the expression in parentheses is true (has a value other than 0). If the parenthesis has the value 0, then the command after else is executed. The else branch does not have to be present, and multiple commands have to be grouped into a block {...}, as shown in the second example.

```
switch (ch=getch())
   {
   case'e':  exit(0);
   case'f':  scanf("%lf",&a); break;
   case's':  scanf("%s",str); break;
   default:  printf("You have entered %c",ch);
             break;
   }
```

Multiple, possibly nested if...else...if...else... instructions can be simplified by replacing them with a switch command. The next character entered on the keyboard is obtained by the function getch(), passed to the variable ch, and then the case instruction that contains the corresponding value ch is executed. If the value is not among those listed, then the default instruction is executed. The last command in each case should be break;. This command causes the program to jump to the end of the switch block {...}.

```
for (sum=0., i=0; i<N; i++) sum +=array[i]; for(;;);
```

We have already encountered the for loop. The first two instructions, separated by a comma, are executed at the beginning. The loop terminates if the value after the first semicolon is zero (i.e., false). The instruction i++ is executed after every pass of the loop. The second example shows a senseless infinite loop, which can only be terminated by interrupting the program execution with Ctrl-c (CONTROL key and c pressed simultaneously).

```
while(ch!='e')
  { ch=getch();
    printf ("The character %c was entered \n",ch);
  }
```

The while loop is another important way of iterating commands. The command (or {...} block) following the while() statement is executed as long as the value of the expression in parentheses is not equal to 0, i.e., as long as it is true. The function getch() reads a character from the keyboard and passes it to the variable ch. Then the character is printed. If the character 'e' is entered the loop stops.

```
double v[6];
double u[5]={1.0, 1.5, 0., 6, -1.};
      v[5]=u[1] * u[0];
```

When declaring vectors, the type and length of the vector must be specified. The indices run from 0 to (length−1) and are referred to with square brackets. Thus v[5] has the value 1.5; v[6] is not defined. Vectors can be initialized while being declared.

```
double mat[6][5];
    for(i=0;i<6;i++)
      {
        for(sum=0., j=0;j<5;j++)
            sum+= mat[i][j]*u[j];
        v[i]=sum;
      }
```

Matrices are declared and referenced with two indices. A separate square bracket is necessary for each index. The first index indicates the row of the matrix, the second indicates the column of the matrix. In the example above, the vector v is the result of the multiplication of the matrix mat by the vector u. The matrix has 6 rows $i = 0, \ldots, 5$ and 5 columns $j = 0, \ldots, 4$. mat contains the address of the first element mat[0][0] and mat[i] that of the ith row, i.e., of the element mat[i][0]. This is important if one wants to pass matrices or rows of matrices to functions.

```
double *a, value, b;
    a = &b;
    scanf ("%lf", a);
    value = *a;
```

There are variables in C which contain the address (memory location) and not the value of a variable. They are declared by **type** *, i.e., the variable a contains the address of a memory location of the given type, or a points to a location of type **double**. Such variables are also called pointers, and the value stored at the corresponding address is referred to by *a. In the example above, the address of b is assigned to a, then a real number is read and stored at this location, and the value stored at a is assigned to **value**. The command scanf("%lf",&value) would have yielded the same result for **value**.

In the following table, x and y are of type **double**. Angles in trigonometric functions are expressed in radians.

sin(x)	Sine of x
cos(x)	Cosine of x
tan(x)	Tangent of x
asin(x)	$\arcsin(x)$ in the range $[-\pi/2, \pi/2]$, $x \in [-1, 1]$
acos(x)	$\arccos(x)$ in the range $[0, \pi]$, $x \in [-1, 1]$
atan(x)	$\arctan(x)$ in the range $[-\pi/2, \pi/2]$
atan2(y,x)	$\arctan(y/x)$ in the range $[-\pi, \pi]$
sinh(x)	Hyperbolic sine of x
cosh(x)	Hyperbolic cosine of x
tanh(x)	Hyperbolic tangent of x
exp(x)	Exponential function e^x
log(x)	Natural logarithm $\ln(x)$, $x > 0$
log10(x)	Logarithm to the base of 10, $\log_{10}(x)$, $x > 0$
pow(x,y)	x^y
sqrt(x)	$\sqrt{x}$, $x \geq 0$
fabs(x)	Absolute value $|x|$
fmod(x,y)	Remainder of x/y

These mathematical functions return a value of type **double**. They have to be declared beforehand with **#include <math.h>**. If the argument is not of type **double**, then it will be changed automatically.

```
double r;
r = rand() / (RAND_MAX +1.);
```

The function **rand()** returns uniformly distributed pseudo-random integers between 0 and **RAND_MAX**. If one needs real-valued random numbers in the interval [0,1], then one can divide **rand()** by **RAND_MAX**. To do this, however, one has to change the denominator to a real number, otherwise the result will be the integer 0. The addition of **RAND_MAX** and the real number 1. yields as its result a number of type **float**. The declaration of the function `rand()` and the definition of the constant `RAND_MAX` are accomplished with `#include <stdlib.h>`.

```
#include <math.h>
main()
{
  double f(double),x;
  for(x=0.;x<10.;x+=.5)
          printf("\n f(%3.1lf) = %4.2e ",x,f(x));
}

double f(double xxx)
{
   double x2;
   x2=xxx*xxx;
   return (x2*exp(-x2/2));
}
```

One can define one's own functions, here $f(x) = x^2 e^{-x^2/2}$. To do so, the function has to be declared in the calling function (here `main()`) and then defined, with the type declarations for the function as well as for the arguments. The commands of the function $f(x)$ are enclosed in braces {...}, and the value of the argument of the `return` statement is returned when execution of the function completes. If no value is returned or no argument is passed, then the type `void` is used. Only the value of the argument `x` is passed, so `f` cannot change the value of the variable `x` here. If, on the other hand, `&x`, the address of `x`, is passed, then `f` can also change the value of `x`:

```
#include <math.h>
main()
{
  double f(double *),x=0.;
  while (x<100.) printf(" \n f(%lf)=%lf",x,f(&x));
}

double f(double *address)
{
  double x2;
  x2=(*address)*(*address);
  *address+=.5;
  return x2*exp(-x2/2.);
}
```

In this case, `double*` indicates a pointer to a variable of type `double`. The transfer of the address instead of the value is necessary if vectors, matrices, or even other functions are to be passed to a function.

```
main()
{
  double u[100],v[100];
  double ScalarProduct(double *, double *, int);
  int n=20,i;
```

```
    for(i=0;i<n;i++) u[i]=v[i]=i;
    printf(" \n %lf \n",ScalarProduct(u,v,n));
}

double ScalarProduct(double u[], double v[], int n)
{
  int i;
  double sum;
  for(i=0,sum=0.;i<n;i++) sum += u[i]*v[i];
  return sum;
}
```

This example shows how vectors are passed to functions. `double u[]` tells the compiler that the function `ScalarProduct` receives the address of the first element of a vector of type `double`; `double *u` is identical to this. No memory location for a vector is reserved in the function; the processing is done with just the address of the first location. `u[3]` counts three positions ahead from this address and returns the value stored at this location. The function does not know how much memory is allocated for the vectors in `main()`; therefore one has to specify their length n.

```
double MatrixFunction(double mat[][10],int rows,
    int columns)
{ ... }
```

With matrices, too, only the address of the first element `&mat[0][0]` is passed, but the program also needs to know the length of the rows (here 10) in order to address each element. Thus the element `mat[i][j]` is at the memory location with the address `&mat[0][0] + i*10 + j`. The number of rows allocated does not have to be known.

```
/* Not the best way */
double u[100],v[100],ScalarProduct(void);int n=20,i=0;
main(){for( ; i<n; i++) u[i]=v[i]=i*i;
printf("%lf",ScalarProduct());}
double ScalarProduct(void){double sum=0.;
for(i=0;i<n;i++)sum+=u[i]*v[i];return sum;}
    /* this is
        a comment */
```

Variables can only be used within the function in which they are defined. If they are to be valid in more than one function, then they have to be declared before those functions outside of any curly braces. `u,v,n,i`, and `ScalarProduct` can thus be used in all functions (and will have the same values everywhere); `sum`, on the other hand, can only be used in `ScalarProduct`. `void` signifies that no variables are passed. Of course, now the function cannot be applied to other vectors x and y; it only acts as an abbreviation for a block of commands.

Attention: Traps!

We want to list some common errors which unfortunately do not produce error messages in many cases.

```
b = mat[i,j];
```

Matrix elements have to be referred to by two indices, each in square brackets, for example mat[i][j]. The command above calls i, then j, and assigns the element mat[j] to the variable b. mat[j], however, is the address of the *j*th row (= &mat[j][0])!

```
if(a=b)   dosomething();
```

The programmer certainly meant to type a==b. The command above assigns b to the variable a and then tests whether the value of a is equal to zero. If not, the function dosomething() is called.

```
integer = 3 * rand() / RAND_MAX;
```

All variables and constants are assumed to be of type **integer**. Some compilers evaluate this expression from right to left. In this case, the quotient and with it the entire expression receives the value zero, since the remainder is lost in the division of two integers. Therefore, one should always place the product in parentheses to obtain the random numbers 0, 1, and 2 regardless of the specific compiler version.

```
int vector[10],i;
for (i=1; i<=10; i++)   vector[i]=i*i;
```

Vectors of length N have indices between i=0 and i=N-1; thus the last command of the for loop, vector[10]=100;, assigns the value 100 to the memory location after the vector without any warning, and the programmer does not know what is being overwritten!

```
main()
{
    char *name;
    printf("What is your name? ");
    scanf( "%s",name);
}
```

name is declared as an address of (= pointer to) a character variable, but the pointer is not initialized anywhere. This means that an unknown number is stored in the variable name which is interpreted as an address; your input is written to this address and destroys other data in the process. The compiler issues a warning about this error, but it does execute it.

```
char   msg[10];
msg ="Hello";
```

Names of vectors are pointers, not variables. "Hello" marks the address of the character string, but the address of the vector msg may not be changed. The following program eliminates both of the above errors

```
main()
{
    char name[20],*msg;
    printf(" \n What is your name? ");
    scanf("%s",name);
    msg="Hello";
    printf("\n %s %s \n", msg,name);
}
```

The difference between "x" and 'x':
The former expression is a character string. It is composed of the two characters 'x' and \0. The latter expression is a single character.

```
int a=100;
scanf("%d",a);
```

scanf interprets a as the memory address number 100 and overwrites whatever was there. The correct read command is scanf("%d",&a);.

Literature

Kernighan B.W., Ritchie D.M. (1988) The C Programming Language. Prentice Hall, Englewood Cliffs, NJ

C. First Steps with Unix

Introduction

This appendix is intended to provide the beginner with a short survey on the most important Unix commands. In addition, we will explain how to access the example programs described in this book over the Internet and how to edit and compile these programs.

Nowadays there are many *public domain* programs for Unix, i.e., programs which may be copied free of charge for noncommercial use. Several of these have become standard programs. Some will be described here, for example less, gzip, gcc, and g++. If these programs are not available to you, you can ask your system manager to install them. These programs are marked by the symbol [*].

Unix is a multiuser and multitasking operating system. Several users can work independently of each other at the same time and several programs can run "simultaneously in the background." "Simultaneous" means that these processes share the computing time.

In order to be able to work on a Unix system one needs a *username* and a *password*. These are obtained from the system manager. For every user there is a separate *home directory* on the hard disk, on which one's own programs can be stored.

After entering the user name and password, you will find yourself either in a *shell* – i.e., the computer waits for commands – or a graphical user interface (usually X11) is started. In the latter case, if a window is not automatically opened, one has to generate a window using the mouse. Then Unix commands can be entered in the window and terminated with RETURN. Windows can be moved with the mouse, can be enlarged or reduced in all directions, or can be placed in the background as an *icon*.

The following groups of commands will be described in this appendix:

- Manipulation of files
- Communication with other computers
- C Compiler

We will not describe the following groups of commands:

- Communication with peripherals (e.g., printing)

236 C. First Steps with Unix

- Monitoring and control of programs running in the background
- Communication with other users (e.g., electronic mail)

Let us point out here that graphical interfaces come with many auxiliary programs (so-called *file managers*) which facilitate file management.

Files

A file is a set of symbols. These can be text data, numbers, or programs. The unit in which the size of files is usually measured is the *byte*. A byte is a number between 0 and 255. Usually a character requires one byte in a file, and a number needs two to four bytes.

Every file has a name. This name can have more than eight letters on almost all systems running under Unix. Capital and lower-case letters are distinguished. Files also have other attributes such as name of the owner, name of the group to which they are assigned, and the rights of the owner, the group, and all other users. For example, one can prevent all other users (except for the system manager) from reading the files.

Files are organized hierarchically in *directories*. These directories can be compared to drawers. The *root* of this directory tree is indicated by the symbol /. Subdirectories are separated by /.

A typical tree structure would be, for example:

```
/  —  /etc    —  /etc/motd
                 /etc/passwd
                 . . .
      /users  —  /users/user1  —  /users/user1/.cshrc
                                  /users/user1/.login
                                  . . .
              . . .
   . . .
```

Here, it is not apparent to the user if the directories are on a local disk or on disks on other computers. The shell remembers a *current directory*. At the beginning, this is the user's home directory. File names that begin with / are absolute, i.e., specified relative to the root. If a file name begins with ~/, then it is taken relative to the user's home directory instead of the root. If it begins with ~*user*/, it is seen in relation to the home directory of *user*. All other file names are interpreted relative to the current directory. One period (.) designates the current directory and two periods (..) the directory above it.

The Most Important Commands

Commands in Unix are composed of the command itself, options, and parameters. Options are designated by the character - preceding them. If several options are used, only one - is necessary. The options have to be placed before the parameters. For example, `ls -a dir` shows all the files of the directory `dir`.

File names can contain *wildcard* characters, which are resolved by the shell. The character * stands for any number of characters. Thus, the command `ls m*.c` displays all files which begin with the letter m and end with .c.

File names are often chosen in such a way that one can recognize the contents of the file, usually by adding a period and a short *extension* to the name. The most important extensions are:

c   C source code

f   Fortran source code

p   Pascal source code (sometimes .pas is used instead)

o   Object file (compiled program code, not yet executable)

m   Text file with *Mathematica* commands

tar A program archive produced with `tar`

gz   A file compressed with `gzip`[(*)]

z   A file compressed with `compress`

Help

By using the command `man`, one can display the description of a command on the screen page by page. Syntax, options, and parameters are explained. For example, with the command `man man` one obtains the description of the command `man` itself.

If one only knows a keyword, but not the exact command, one can use `man -k keyword` to obtain a list of the commands which contain the word `keyword` in their one-line description.

File Manipulation

A file can easily be created or modified with a so-called editor. This will be described later. Text files can be displayed with `cat`, `more`, or `less`[(*)]. `more` displays the file page by page; by using `less` one can move around in the text at will by pressing a key.

Copying a file can be done with the Unix command `cp` (*copy*). The syntax is `cp input output`. The file `input` is copied to the file `output`. In order to copy a directory, the option `-R` has to be specified. By using this, the files contained in the directory and its subdirectories are copied recursively.

The command mv (*move*), which makes it possible to rename files or to move them between different directories, has the same syntax: mv old new renames the file old to new. This can also be done with directories.

A file can be deleted with the command rm (*remove*). rm file deletes the file file. The file cannot be restored after that. The command rm *.txt removes all files with the extension txt in the current directory. Entire directories, including their files, can be deleted recursively with the command rm -R. One should be very careful with this command to avoid deleting important data. In the beginning this should only be used in conjunction with the option -i (*interactive*). Then the computer will explicitly ask, for each file, if it should really be deleted. So for example one should type rm -Ri directory.

Information about Directories, Current Directory

The name of the current directory can be shown with pwd (*present working directory*). All file names which do not begin with / are interpreted relative to this current directory, as mentioned above. The command cd dir sets the current directory to dir. If the parameter is omitted, then the user's home directory is used.

Directories can be created with mkdir (*make directory*) and deleted with the command rmdir (*remove directory*); the command mkdir temp creates the directory temp, the command rmdir temp deletes the empty (!) directory temp. Directories which contain files can only be deleted with rm -R.

The command ls (*list*) is used to display the contents of a directory. With ls dir the contents of the directory dir are displayed. The option -a causes files whose names begin with a period to be shown too, for example the file .cshrc, which is processed when starting the shell. The option -l brings up the list in a long format, with information about access rights, users, groups, date of last modification, and size.

Password

The user's own password can be changed with the command passwd. To do this, one has to type in the old password, then the new one, and then the new one again to be sure that it has not been mistyped. The password does not appear on the monitor.

NIS (Network Information Service) – previously *Yellow Pages* – is often used if one works on several computers networked together. Then there is just one password file for all (networked) computers. In that case there is a *server* which manages the passwords, and the command for changing the password is yppasswd.

Rerouting Input and Output

The shell, i.e., the command interpreter, allows one to reroute inputs and outputs by inserting the characters < and > respectively. The output of a program turns into the input of another program via the symbol | (*pipe*). For example, ls dir | more displays the contents of the directory dir page by page. The command ls dir > temp generates the file temp and writes the output of the command ls to it.

The Standard Editor vi

Several so-called editors are at the user's disposal for creating and modifying files. vi (*visual*) is the most widely used. It has a lot of possibilities, but it is not easy to use, at least not in the beginning. The editor emacs, by the GNU consortium, is used frequently as well. Here we just want to discuss vi, since it comes as part of every Unix system.

The editor is started with the command vi file. This causes the file file to be loaded (if it already exists) or newly created. When working with vi, there are three different modes:

- the command mode
- the insert mode
- the command line mode

At startup, the editor is in command mode. This means that the keyboard inputs are interpreted as commands. Important vi commands in this mode are:

A	Jumps to the end of the line and switches to insert mode (*Append*)
I	Jumps to the beginning of the line and switches to insert mode (*Insert*)
i	Switches to insert mode (*Insert*)
Esc	Switches back from insert mode to command mode
J	Appends the next line to the current line (*Join*)
x	Deletes the character at the cursor position
dd	Deletes the current line
D	Deletes from the cursor position to the end of the line
dw	Deletes one word
nG	Jumps to line n
G	Jumps to the last line
/*pattern*	Searches for *pattern*

A colon : is used to switch to command line mode. The most important commands in that mode are:

:x	Saves the document and leaves the editor (*Exit*)
:q	Leaves the editor if no changes have been made (*Quit*)
:q!	Leaves the editor, discarding changes (*Quit*)
:w	Saves the document (*Write*)
:wq	Saves the document and leaves the editor
:r *file*	Reads the file called *file* (*Read*)
:set nu	Switches on line numbering (*Set Numbers*)

There are many more commands. For example, one can of course mark blocks of text and copy them, or indent entire blocks.

The C Compiler

The translation of a C program into executable code is done in two steps. In the first step the program is translated (or *compiled*) into an object file. The latter then contains the compiled program. In the second step, this object is *linked* with standard libraries and possibly other object files to form an executable program.

The C compiler which comes with a Unix system is called cc. With the call cc file.c the program file.c is compiled and then immediately linked. The finished executable program is called a.out. If mathematics routines, e.g., trigonometric functions, are to be used, the mathematics library has to be linked to the program. This is done via the option -lm.

Important options are:

-o name	The output file is called name rather than a.out
-O	Optimizes the code
-c	Generates the object file only
-I dir	Searches for include files in dir, in addition to the standard search path
-lx	The linker links the library lib$x$ to the program
-L dir	The linker searches dir for program libraries, in addition to the standard search path

Another C compiler, which is available at no cost, is GNU's gcc[*]. It often produces faster code than the system compiler. Furthermore it supports C++ via the call g++.

Make

In order to compile larger program packages, the auxiliary program make is often used. It reads a file called Makefile and processes the instructions given there. We will not discuss the format here.

Networks

Many Unix computers are connected via the *Internet*. In the remainder of this section we will explain the most important commands related to this network.

Every computer connected to the Internet has an address. This address consists of four numbers, each of which can vary between 0 and 255. Every address has a name associated with it. ftp.physik.uni-wuerzburg.de, for example, currently has the address 132.187.40.15. The individual groups of numbers cannot be mapped to the components of the name in a unique way. For example, any address at the University of Würzburg (.uni-wuerzburg.de) starts with 132.187, but the physics department uses additional subdomains other than 40. Since the numbers may change if a network is reorganized, one should use the names whenever possible.

Connections to Other Computers

The command telnet computer can be used to establish a connection to another computer on the Internet. Here computer can be either the computer's name or its Internet address.

Typing telnet ftp.physik.uni-wuerzburg.de results in the following output if there is a physical connection and our computer is running:

```
Trying...
Connected to wptx15.physik.uni-wuerzburg.de.
Escape character is '^]'.

(some machine-specific informations)

login:
```

At this point, the user has to enter his or her username and password.

One should always close telnet sessions correctly, i.e., leave the computer via logout. If nothing else works, one can get to a telnet menu by using the *escape character* that is specified. The symbol ^] used here means that the CONTROL key and] have to be pressed simultaneously. Once in the telnet menu, one can get short instructions via the command ?. The connection can be terminated with the command close.

A possibility of working on a different Unix computer is provided by the command rlogin. This command assumes the same username, which can be changed explicitly by the option -l.

Copying Programs from Other Computers

In order to exchange files between computers there is FTP, a *file transfer protocol*. A connection is established via `ftp computer`. Again, `computer` can be either an Internet address or a name. Here, too, a user has to identify him- or herself; the computer asks for username and password. After a successful `login` the following commands can be used, among others:

`bi`	sets the transfer mode to *binary*, i.e., the data are transferred without change
`as`	sets the transfer mode to *ASCII* (text mode). In this mode, control characters may be suppressed. There may also be a system-dependent conversion of the characters
`ls`	lists the files in the remote directory (*list*)
`cd dir`	changes the remote directory to `dir`
`get file`	copies the file `file` from the remote computer to the local one
`put file`	copies the file `file` from the local computer to the remote one
`mget files`	similar to `get`, but this command accepts wildcard characters in the file name and transfers multiple files
`mput files`	similar to `put`, but this command accepts wildcard characters in the file name and transfers multiple files

Internet File Transfer

Many institutions offer programs which anyone may copy freely. These can be obtained through the Internet, using FTP. To this end, special *FTP servers* offer the option of logging in under the username `ftp` or `anonymous`. The password required is one's own e-mail address. For example, the programs described in this book are available from the FTP server

```
ftp.physik.uni-wuerzburg.de
```

(current IP address: 132.187.40.15) in the directory `/pub/cphys`.

Data Compression

In order to archive data it is useful to store several programs together under one name. This is frequently done with the program `tar` (*tape archiver*). The command `tar cf file.tar dir` is used to archive the directory `dir` with all its files and subdirectories in the single file `file.tar`. With the command `tar tf file.tar` one obtains a listing of the contents of the archive, and with the command `tar xf file.tar` the directory is restored.

Then, in order to save disk space, compression programs are used. One frequently used program is called gzip[*]. The command gzip file compresses the file file. The output file has the name file.gz. To decompress, one can use either the command gzip -d file.gz or gunzip file.gz.

X11 across a Network

It is no problem to run X11 applications across a network, i.e., to do the calculations on one computer and display the output on another. Let us assume we want to start an X program on the computer remote and have it display its windows on the X11-capable monitor of our local computer local. First of all, we have to tell our computer to accept displays from the other one. This is done via xhost + remote. On the computer remote we have to set the environment variable DISPLAY. Both in csh and in tcsh this is done via setenv DISPLAY local:0. The symbol :0 represents the first X session. All X programs called from now on will send their graphics output to the monitor of local.

Literature

Christian K., Richter S. (1994) The Unix Operating System. Wiley, New York

D. First Steps with *Xgraphics*

In our sample programs, we frequently calculate the dynamics of certain systems which are displayed on the screen immediately after the data become available. Such a graphical representation of results calculated in a C program is not easy on a workstation running under the Unix operating system. While the user interface X-Windows, which can be used to program elementary graphics commands, is available on almost all computers, its use is very complicated for beginners, as even simple programs require several pages of source code.

In order to simplify the writing of graphics programs, Martin Lüders has developed *Xgraphics*. It exclusively uses the routines from Xlib, the standard library of the X-Windows system, and provides simple commands for managing windows and drawing in them. The many parameters which have to be passed to the Xlib routines are managed internally, so only the really essential parameters have to be specified in the program. Most importantly, though, the user is freed from a large part of the window management. For example, in simple cases one does not have to worry about what happens if a window is enlarged by the user or reduced to an icon. This allows us to actually place the focus of the programs on the physics. Still, the data structures of Xlib are immediately accessible, so one can add Xlib commands or extensions to Xlib at any time.

The drawing regions, called *Worlds*, are an important building block of *Xgraphics*. They allow the definition of a local coordinate system which is adapted to the problem to be programmed rather than having to adjust to the present size of the current window. The most important commands of *Xgraphics* are: create a window (`CreateWindow`), create a drawing region with user-defined coordinates in the window (`CreateWorld`), create control buttons that can be accessed by mouse clicks (`InitButtons`), react to movements of the mouse or to keystrokes (`GetEvent`), read a number (`GetNumber`), draw a point in pixel (`DrawPoint`) or user-defined coordinates (`WDrawPoint`), draw a circle (`DrawCircle` or `WDrawCircle` respectively), write text (`DrawString`, `WDrawString`), etc. The source code of the C program `Xgraphics.c` has to be compiled together with the user's program. In doing this, the X11 library and the appropriate paths have to be specified. This is done by the program `compile`. In the user program, the header file `Xgraphics.h` has to be included via `#include`. The postscript file `Xgraphics.ps` provides a detailed description.

The program package, including the documentation, some demonstration programs, and the X-Windows versions of the sample programs from this book, can be found on the enclosed CD-ROM. They have been combined into groups by using the Unix command `tar`. The archives are

- `xgraphics.tar`   The actual *Xgraphics* package and the documentation
- `demo.tar`   Some demonstration programs
- `physics.tar`   The examples from the book

The archives are unpacked with the command

```
tar xf <file>.tar
```

In addition, you can copy them to your computer via the Internet. This is done as follows:

```
ftp  ftp.physik.uni-wuerzburg.de
```

The computer asks for the username:

```
ftp
```

The computer asks for a password:

```
reader@yourhost.yourdomain
```

The correct directory has to be selected:

```
cd pub/Xgraphics
```

Now all files are copied:

```
mget *
```

For each file in turn, the computer asks whether it should be copied (answer y or n). Finally, the connection is closed:

```
bye
```

In addition, there are some pages about *Xgraphics* on the World Wide Web, including the complete documentation and pointers to new versions, if applicable. The web address is

```
http://theorie.physik.uni-wuerzburg.de/Xgraphics/
```

We want to use a simple example to demonstrate that just a few commands are sufficient to display a physics problem on the screen. The example is the random motion of a particle (*random walk*), for which there is a well-developed mathematical description. In the model a particle randomly jumps from its current position to one of the four adjacent sites on a square lattice. This is done in the following C program:

```
#include <stdlib.h>
#include <math.h>
#include "Xgraphics.h"

#define MAXX 640
#define MAXY 480
#define N 250000

main()
{
    int x=MAXX/2,y=MAXY/2,r,i;

    InitX();
    mywindow=CreateWindow(MAXX,MAXY,"random walk");
    ShowWindow(mywindow);

        for(i=0;i<N;i++)
            {   DrawPoint(mywindow,x,y,1);
                r=rand()/(RAND_MAX+1.)*4.;
                switch(r)
                        {   case 0: x++;break;
                            case 1: x--; break;
                            case 2: y++;break;
                            case 3: y--;break;          }
            }
    getchar();
    ExitX();
}
```

Here x and y are the pixel coordinates of the particle in the window, whose size is determined by MAXX and MAXY. In each step, one of these coordinates is increased or decreased by one according to the random number $r \in \{0, 1, 2, 3\}$. Thus five graphics commands are enough to generate interesting movements on the screen. The program, called walk.c, is compiled via compile walk and executed by the command walk. Before the compilation, the appropriate path names for one's own computer may have to be specified in compile, and the file has to be made executable via chmod u+x compile. The result should be the picture shown in Fig. D.1. Then if one enters any character in the window from which the program was launched, the picture of the random walk disappears.

Additional demonstration programs, which also show the other capabilities of *Xgraphics*, can be found on the enclosed CD-ROM.

Literature

Lüders M. (1997) Introduction to Xgraphics. Postscript file on the enclosed CD-ROM; also available via FTP from ftp.physik.uni-wuerzburg.de, file Xgraphics.ps in the directory /pub/Xgraphics/

Jones O. (1989) Introduction to the X Window System. Prentice Hall, Englewood Cliffs, NJ

Fig. D.1. Random walk consisting
of 250 000 steps on a square lattice

E. Program Listings

In this appendix we print the listings of some of our sample programs, namely almost all *Mathematica* programs and a few of the C programs for the PC. Unfortunately we cannot print all source codes, owing to limitations of space, but all programs are available on the enclosed CD-ROM and can be printed. The names of the *Mathematica* programs have the form **name.m** throughout. The PC versions of the C programs, **name.c**, have been compiled with *Turbo C* by Borland; the resulting executable files, **name.exe**, are available on the enclosed CD-ROM.

Some of the C programs can be controlled by keyboard inputs which are indicated on the screen in the line starting with the word *Commands*. The first letter of each keyword listed is the key that controls the algorithm; for example, *exit* means that the key e terminates the program.

The CD-ROM also contains versions of all sample programs for Unix or Linux. In addition, the files can be copied (possibly in an improved version) via the Internet from the FTP server of the Institute for Theoretical Physics of the University of Würzburg (see Appendix C). The relevant directories on this server are:

```
/pub/cphys/dos/
/pub/cphys/mathematica/
/pub/cphys/unix/
/pub/Xgraphics/
```

Additional information about the programs can be found on the World Wide Web under the address

```
http://theorie.physik.uni-wuerzburg.de/cphys/
```

1.1 Function versus Procedure: sum.m and sum.c

```
Print["\n Function versus Procedure \n"]
dataset=Table[Random[],{10000}]
average[data_,length_]:=Block[{sum,average},
                              sum=0.;
                              Do[sum=sum+data[[i]],{i,length}];
                              average=sum/length
                              ]
average[data_]:=Apply[Plus,data]/Length[data]
```

sum.c

```c
#include <stdlib.h>
#include <time.h>

main()
{
  float average(float *,int);
  int i;
  clock_t  start,end;
  float dataset[10000];
  clrscr();

  for (i=0;i<10000;i++) dataset[i]=random(1000)/1000.;

  printf( "\n Ten thousand random numbers are added;\n"
    "this is repeated a hundred times, with the result:\n");

  start=clock();
  for(i=0;i<100;i++) average(dataset,10000);
  printf( "          average = %f\n",average(dataset,10000));

  end=clock();
  printf( "          time= %f  sec ",(end-start)/CLK_TCK);
  getch();
}

float average( float* data,int n )
{
  float sum=0. ;
  int i;
  for(i=0;i<n;i++) sum=sum+data[i] ;
  return sum/n;
}
```

1.2 Nonlinear Pendulum: pendulum.m

```
Print["\n Nonlinear Pendulum \n"]
T[phi0_]=4 EllipticK[Sin[phi0/2]^2]
plot1:=Plot[T[phi0],{phi0,0,3.14},PlotRange->{0,30},
          Frame -> True, FrameLabel->{"phi0","T"}]

sinuspsi[t_,phi0_]=JacobiSN[t,Sin[phi0/2]^2]
phinorm[x_,phi0_]=2 ArcSin[Sin[phi0/2]*
                           sinuspsi[x T[phi0],phi0]]/phi0
phi0[1]=N[.1 Pi]
phi0[2]=N[.8 Pi]
phi0[3]=N[.95 Pi]
phi0[4]=N[.99 Pi]
phi0[5]=N[.999 Pi]
flist=Table[phinorm[x,phi0[i]],{i,5}]
plot2:=Plot[Evaluate[flist],{x,0.,1.},
          Frame -> True,
```

```
                FrameLabel->{"t/T","phi/phi0"},
                PlotStyle -> Thickness[0.001]]

list:=Table[phinorm[x,N[.999 Pi]],{x,0,.99,.01}]
foulist:=Take[Abs[Fourier[list]],15]
plot3:= (gr1 = ListPlot[foulist,PlotRange->{{0,15},{-1,8}},
                      Frame -> True,
                      PlotStyle -> PointSize[0.02],
                      FrameLabel->{"s","Abs[b(s)]"},
                      DisplayFunction -> Identity] ;
              Show[gr1, Graphics[Line[{{0,0},{0,15}}]],
                      DisplayFunction -> $DisplayFunction ] )

f=1/Sqrt[1 - m Sin[psi]^2]
g:=Series[f,{m,0,10}]
tseries:=4 Integrate[g,{psi,0,Pi/2}] /. m->Sin[phi0/2]^2

e=phidot^2/2-Cos[phi]
plot4:=ContourPlot[e,{phi,-Pi,Pi},{phidot,-3,3},
                   Contours->{-.5,0,.5,1,1.5,2},
                   ContourShading->False,PlotPoints->100,
                   FrameLabel -> {"phi", "phidot"}]
```

1.3 Fourier Transformations: fourier.m

```
Print["\n Fourier transformation"]
f[t_]=(Sign[1-t]+Sign[1+t])/2
fs[w_] = Integrate[Exp[I*w*t], {t, -1, 1}]/T // ComplexExpand
T=10
SetOptions[ListPlot,Frame->True]
SetOptions[Plot,Frame->True,RotateLabel->True]
plot1:=Plot[fs[w],{w,-5Pi,5Pi}, PlotRange -> All]

p2list:= p2list = Table[ Abs[ fs[ k*2 Pi/T] ],{k,64}]
plot2:=(g1=ListPlot[p2list,PlotJoined->True,PlotRange->All,
                   PlotStyle -> Thickness[0.0],
                   DisplayFunction -> Identity];
       g2=ListPlot[p2list,PlotRange->All,
                   PlotStyle -> PointSize[0.015],
                   DisplayFunction -> Identity];
          Show[g1,g2,DisplayFunction -> $DisplayFunction] )

flist := flist = Table[N[f[-5+10 r/64]],{r,64}]
fslist := fslist = Fourier[flist]/Sqrt[64]
p3list:= p3list = Abs[fslist ]
plot3:=(g1=ListPlot[p3list, PlotJoined->True,PlotRange->All,
               PlotLabel->"Discrete Fourier transformation",
                   PlotStyle -> Thickness[0.0],
                   DisplayFunction -> Identity];
       g2=ListPlot[p3list,PlotRange->All,
                   PlotStyle -> PointSize[0.015],
                   DisplayFunction -> Identity];
          Show[g1,g2, DisplayFunction -> $DisplayFunction] )
```

```
fapp1[t_]=Sum[fslist[[s]] Exp[-2 Pi I /64 (s-1)(64t/T+31)],
                {s,64}]//N
plot4:=Plot[{Re[fapp1[t]],f[t]},{t,-5,5}]
fapp2[t_]=Sum[N[fslist[[s]]*
                    Exp[-2 Pi I /64 (s-1)(64t/T+31)]],{s,32}]+
            Sum[N[Conjugate[fslist[[34-s]]]*
                    Exp[-2 Pi I /64 (s-33)(64t/T+31)]],{s,32}]

plot5:=Plot[{Re[fapp2[t]],f[t]},{t,-5,5}]
```

1.4 Smoothing of Data: smooth.m

```
Print["\n Smoothing Data by a Convolution \n"]
data = Table[N[BesselJ[1,x]+.2 (Random[]-1/2)],{x,0,10,10./255}]
xdata = Range[0,10,10./255]
SetOptions[ListPlot,Frame->True]
SetOptions[Plot,Frame->True,RotateLabel->True]
plot1 := p1 = ListPlot[Thread[Join[{xdata},{data}]],
                        PlotStyle->PointSize[.01]]

sigma= 0.4
kernel = Table[ N[ Exp[-x^2/(2*sigma^2)]],{x,-5,5,10./255} ]
kernel = RotateLeft[kernel,127]
kernel = kernel/Apply[Plus,kernel]
plot2:=p2=ListPlot[Thread[Join[{xdata},{kernel}]],PlotRange->All]

smooth=Sqrt[256] InverseFourier[Fourier[data] Fourier[kernel]]
smooth=smooth//Chop
plot3:=p3=ListPlot[Thread[Join[{xdata},{smooth}]]]

plot4:=p4=Plot[BesselJ[1,x],{x,0,10},PlotStyle ->Thickness[0.001]]

plot5:= (If[Head[p1]==Symbol,plot1];If[Head[p3]==Symbol,plot3];
          If[Head[p4]==Symbol,plot4];Show[p1,p3,p4])
```

1.5 Nonlinear Fit: chi2.m

```
Print["\n Nonlinear Fit \n"]
Needs["Statistics`Master`"]
SeedRandom[12345]
data = Table[{t,Sin[t] Exp[-t/10.]+.4 Random[]-.2}//N,
                {t,0,3Pi,.3 Pi}]
SetOptions[ListPlot,Frame->True]
SetOptions[Plot,Frame->True,RotateLabel->True]
plot1:= (gr1 = ListPlot[data,PlotStyle -> PointSize[0.02],
                        DisplayFunction -> Identity];
          gr2 = Plot[Sin[t] Exp[-t/10.],{t,0,3Pi},
                        DisplayFunction -> Identity];
          Show[gr1,gr2, DisplayFunction -> $DisplayFunction])

f[t_]=a Sin[om t + phi] Exp[-t b]
```

```
sigma2=NIntegrate[x^2,{x,-.2,.2}]/.4
sigma = Sqrt[sigma2]
square[{t_,y_}]=(y-f[t])^2/sigma2
chi2=Apply[Plus,Map[square,data]]
find:=FindMinimum[chi2,{a,0.9},{om,1.1},{phi,0.1},{b,.2}]
fit:=If[$VersionNumber>=3.,
        BestFitParameters /. NonlinearRegress[data,f[t],t,
                {{a,1.1},{om,1.1},{phi,.1},{b,.2}},
                ShowProgress->True],
        NonlinearFit[data,f[t],t,
                {{a,1.1},{om,1.1},{phi,.1},{b,.2}},
                ShowProgress->True]]
pvalue[x_]=1.-CDF[ChiSquareDistribution[7],x]
interval={Quantile[ChiSquareDistribution[7],.05],
          Quantile[ChiSquareDistribution[7],.95]}
lim[x_]=Quantile[ChiSquareDistribution[4],x]
plot2:=Plot[PDF[ChiSquareDistribution[7],x],{x,0,20}]


rule = If[$VersionNumber>=3.,
        BestFitParameters /. NonlinearRegress[data,f[t],t,
                {{a,1.1},{om,1.1},{phi,.1},{b,.2}}],
        NonlinearFit[data,f[t],t,
                {{a,1.1},{om,1.1},{phi,.1},{b,.2}}]]
a0=a/.rule; b0=b/.rule; om0=om/.rule; phi0=phi/.rule;
chi2min = chi2/.rule
plot3:=ContourPlot[chi2/.{phi->phi0,om->om0},
        {a,.4,1.3},{b,-.04,.2},
        ContourShading->False,
        Contours->{chi2min+lim[.683],chi2min+lim[.9]},
        PlotPoints->50,FrameLabel->{"a","b"}]

plot4:=ContourPlot[chi2/.{phi->phi0,a->a0},
        {om,.90,1.05},{b,0.0,.15},
        ContourShading->False,
        Contours->{chi2min+lim[.683],chi2min+lim[.9]},
        PlotPoints->50,FrameLabel->{"om","b"}]

g[t_]= f[t]/.rule
step:=(data2=N[Table[{t,g[t] + .4 Random[]-.2},
                {t,0,3Pi,.3 Pi}]];
      If[$VersionNumber>=3.,
        BestFitParameters /. NonlinearRegress[data2,f[t],t,
                {{a,a0},{om,om0},{phi,phi0},{b,b0}}],
        NonlinearFit[data2,f[t],t,
                {{a,a0},{om,om0},{phi,phi0},{b,b0}}]] )
tab:= tab = Table[{chi2,a,b,om,phi}/.step,{100}]
abtab := Map[Take[#,{2,3}]&,tab]
plot5:= ListPlot[abtab,PlotRange->{{.4,1.2},{-0.045,.2}},
                AspectRatio -> 1, Frame -> True,
                FrameLabel ->{"a","b"}, Axes -> None,
                PlotStyle -> PointSize[0.015]]
```

1.6 Multipole Expansion: multipol.m

```
Print["\n Multipoles for a Static Potential \n "]
SeedRandom[123456789]
 point:={2Random[]-1,2Random[]-1,0}
Do[r[i]=rpoint,{i,10}]
p1=Graphics[Table[Line[{Drop[r[i],-1]-{0.08,0},
                        Drop[r[i],-1]+{0.08,0}}],{i,5}]]
p2=Graphics[Table[Line[{Drop[r[i],-1]+{0,0.08},
                        Drop[r[i],-1]-{0,0.08}}],{i,5}]]
p3=Graphics[Table[Line[{Drop[r[i+5],-1]-{0.08,0},
                        Drop[r[i+5],-1]+{0.08,0}}],{i,5}]]
p4=Graphics[{Thickness[0.001],
             Table[Circle[Drop[r[i], -1], 0.1],{i,10}]}]
SetOptions[ListPlot,Frame->True]
SetOptions[Plot,Frame->True,RotateLabel->True]
plot1:=Show[p1, p2, p3, p4, Frame ->True,
            AspectRatio -> Automatic,
            PlotRange -> {{-2, 2}, {-2, 2}},
            FrameLabel -> {"x", "y"}]

dist[r_,s_]=Sqrt[(r-s).(r-s)]
pot[rh_]:=Sum[ 1/dist[rh,r[i]]-1/dist[rh,r[i+5]] ,{i,5}]
plot2:=Plot3D[pot[{x,y,0}] ,{x,-2,2},{y,-2,2},
            AxesLabel->{"x","y","Phi"},
            PlotRange -> {-8,8},
            PlotPoints->40]

plot3:=ContourPlot[pot[{x,y,0}] ,{x,-2,2},{y,-2,2},
            PlotPoints->40,
            ContourShading->False,AxesLabel->{"x","y"} ]

quadrupole[r_]:=Table[3 r[[k]] r[[l]] - If[k==1,r.r,0],
                    {k,3},{l,3}]
qsum=Sum[quadrupole[r[i]]-quadrupole[r[i+5]],{i,5}]
magnitude[r_]=Sqrt[r.r]
dipole=Sum[r[i]-r[i+5],{i,5}]
pot1[r_] = dipole.r/magnitude[r]^3
pot2[r_] = pot1[r] + 1/2/magnitude[r]^5 * r.qsum.r
path={.6,y,0}
plot4:=Plot[{ pot1[path],pot2[path],pot[path]},{y,-2,2},
            Frame -> True,
            FrameLabel -> {"{.6,y,0}","potential"},
            PlotStyle -> {Dashing[{.01,.01}],
                        Dashing[{.03,.03}],Dashing[{1.0,0}]}]

Needs["Graphics`PlotField`"]
efield=-{D[pot[{x,y,0}],x],D[pot[{x,y,0}],y]}
direction=efield/magnitude[efield]
plot5:=PlotVectorField[direction,{x,-10,10},{y,-10,10},
                    PlotPoints->30]
```

1.7 Line Integrals: lineint.m

```
Print["\n Line Integrals \n"]
r1={Cos[2Pi t],Sin[2Pi t],t}
r2={1,0,t}
r3={1-Sin[Pi t]/2,0,1/2(1-Cos[Pi t])}
p1:=ParametricPlot3D[Evaluate[r1],{t,0,1},
                     DisplayFunction->Identity]
p2:=ParametricPlot3D[Evaluate[r2],{t,0,1},
                     DisplayFunction->Identity]
p3:=ParametricPlot3D[Evaluate[r3],{t,0,1},
                     DisplayFunction->Identity]
plot1:=Show[p1,p2,p3,PlotLabel->" Line integrals",
                AxesLabel->{"x","y","z"},
                DisplayFunction->$DisplayFunction]

f[{x_,y_,z_}]={2 x y +z^3,x^2,3 x z^2}
v[r_]:=D[r,t]
int[r_]:=Integrate[f[r].v[r],{t,0,1}]
r4=t{x,y,z}
curl[{fx_,fy_,fz_}]:={ D[fz,y]-D[fy,z],
                      D[fx,z]-D[fz,x],
                      D[fy,x]-D[fx,y]  }
```

1.8 Maxwell Construction: maxwell.m

```
Print ["\n Maxwell Construction for the van der Waals Gas \n"]
pp=t/(v-b)-a/v^2
eq1=D[pp,v]==0
eq2=D[pp,{v,2}]==0
sol=Solve[{eq1,eq2},{t,v}]
pc=pp/.sol[[1]]
critical = Join[sol[[1]],{Rule[p,pc]}]

p[v_] = 8 t/(3 v-1)-3/v^2
Off[Integrate::gener]
eq3 = p[v1]==p[v3]
eq4 = p[v1]*(v3-v1)==Integrate[p[v],{v,v1,v3}]
On[Integrate::gener]
pmax[v_]:=If[v < vleft || v > vright, p[v], p[vleft]]
plot[T_]:= Block[{}, t = T;
                If[ t >= 1,
                Plot[p[v], {v, .34, 5.},
                    PlotRange -> {{0,5},{0,2}},
                    Frame -> True,
                    FrameLabel -> {"v","p"}],
           vmin = v /. FindMinimum[ p[v], {v,0.4}][[2]];
           vmax = v /. FindMinimum[-p[v], {v,1.}][[2]];
           vtest = (vmin + vmax)/2 ;
           r = Solve[ p[v] == p[vtest], v];
           v1start = v /. r[[1]];
           v3start = v /. r[[3]];
           frs = Chop[FindRoot[{eq3,eq4},{v1,v1start},
```

```
                                {v3,v3start}]];
                  vleft=v1/.frs; vright=v3/.frs;
                  Plot[{pmax[v],p[v]},{v,0.34,5.},
                       PlotRange->{{0,5},{0,2}},
                       Frame -> True, FrameLabel -> {"v","p"} ]]]
```

1.9 Best Game Strategy: game.c

Available on CD-ROM only.

2.1 Quantum Oscillator: quantumo.m

```
Print["\n Quantum Oscillator \n"]
q[j_,k_]:= Sqrt[(j+k+1)]/2 /; Abs[j-k]==1
q[j_,k_]:= 0 /; Abs[j-k]!= 1
q[n_]:= Table[q[j,k], {j,0,n-1}, {k,0,n-1}]
h0[n_]:= DiagonalMatrix[Table[i+1/2,{i,0,n-1}]]
h[n_]:= h0[n] + lambda q[n].q[n].q[n].q[n]
ev[n_]:= Eigenvalues[h[n]]
plot1:= Plot[Evaluate[ev[4]], {lambda, 0, 1},
              Frame -> True,
              FrameTicks -> {Automatic,
         Table[{0.5*j,If[EvenQ[j],ToString[j/2],""]},{j,19}]},
              FrameLabel -> {"lambda","energy"},
              PlotLabel ->"Eigenvalues of h[4]" ]

evnum[n_,la_]:= Sort[Eigenvalues[N[h[n]/. lambda->la]]]
evdimlist[n_,la_]:= Table[{1/i, evnum[i,la][[1]]}, {i,7,n}]
plot2:=( gr1=ListPlot[evdimlist[20,.1],
                        Axes -> None, Frame -> True,
                        PlotRange -> All,
                        FrameLabel -> {"1/n","E0(n)"},
                        PlotStyle -> PointSize[0.02],
                        DisplayFunction -> Identity] ;
        gr2 = Graphics[{Thickness[0.001],
                        Line[{{1/20,0.559146327},
                              {1/7 ,0.559146327}}]}];
        Show[gr1,gr2,DisplayFunction -> $DisplayFunction] )

evlist:= evlist= Table[{la,evnum[20,la]},{la,0,1,.05}]
ev2[k_]:=Table[{evlist[[i,1]],evlist[[i,2,k]]},
               {i,Length[evlist]}]
plot3:= ( Table[gr[k]=ListPlot[ev2[k],PlotJoined ->True,
                        DisplayFunction -> Identity,
                        Frame -> True,
                        FrameLabel -> {"lambda","energy"},
                        FrameTicks -> {Automatic,
                  Table[{0.5*j,If[EvenQ[j],ToString[j/2],""]},
                        {j,21}]}],{k,5}];
          Show[gr[1],gr[2],gr[3],gr[4],gr[5],
               DisplayFunction -> $DisplayFunction] )
```

2.2 Electrical Network: network.m

```
Print["\n Electric Circuits \n"]
eq1={vr+vo==1,ir==ic+il,vr==ir r,
     vo==ic/(I omega c),vo==I omega l il}
sol=Solve[eq1,{vo,vr,ir,ic,il}][[1]]
vos=(vo/.sol)//Simplify
numvalue = {c -> N[10^(-6)], l -> N[10^(-3)]}
vosnum = vos /. numvalue
flist=Table[vosnum /.r->100.0*3^s, {s,0,3}]
plot1:=Plot[Evaluate[Abs[flist]],{omega,20000,43246},
            PlotRange->{0,1}, Frame -> True,
            FrameLabel -> {"omega","Abs[vo]"},
            FrameTicks -> {{20000,30000,40000},Automatic} ]

plot2:=Plot[Evaluate[Arg[flist]],{omega,20000,43246},
            Frame -> True,
            FrameLabel -> {"omega","Phase(vo)"},
            FrameTicks -> {{20000,30000,40000},Automatic} ]

vsaw[t_] := -1+2 t/T
omegares=1/Sqrt[l c] /. numvalue //N
alist=Table[N[vsaw[(n-1)/256 T]],{n,256}]
blist=InverseFourier[alist]
plot3[fac_,w_]:=Block[{volist,vtrans,omegai,plotlist},
                omegai = fac*omegares;
                volist=Join[
                  Table[vosnum/.{omega->omegai*(s-1),r->w},
                        {s,128}],{0.0},
                  Table[vosnum/.{omega->omegai*(s-128),r->w},
                        {s,1,127}] ];
              vtrans=Fourier[blist volist]//Chop;
              plotlist=Table[{k/256,vtrans[[Mod[k-1,256]+1]]},
                             {k,768}];
              ListPlot[plotlist,
                       Frame -> True,
                       FrameLabel->{"t/T","vo(t)"},
                       PlotRange -> All]]

eq2={ir(r + 1/(I omega c) + I omega l) + vo ==1,
     ir == (I omega c + 1/(I omega l)) vo }
sol2=Solve[eq2,{vo,ir}][[1]]
vos2=(vo/.sol2)//Simplify
irs = Simplify[ir /. sol2]
power = r^2 Abs[irs]^2
flist3=Table[Abs[vos2] /. numvalue /. r->10*3^s ,{s,0,2}]
plot4:=Plot[Evaluate[flist3],{omega,10000,70000},
        PlotRange->All,Frame -> True,
        PlotStyle->{Thickness[0.006],Dashing[{.01,.008}],
                    Thickness[0.003]},
        FrameTicks -> {{10000,30000,50000,70000},Automatic},
        FrameLabel ->{"omega","|vo|"}]

plot5:=Plot[(power/.numvalue)/.r->10.0,{omega,10000,70000},
```

```
                PlotRange->All,Frame -> True,
                FrameTicks -> {{10000,30000,50000,70000},
                                  Automatic},
                FrameLabel ->{"omega","P/P0"}]
```

2.3 Chain Vibrations: chain.m

```
Print["\n Oscillator Chain \n "]
mat1 = { { 2f           , -f , 0   , -f*Exp[-I q] },
         { -f           , 2f , -f  , 0            },
         { 0            , -f , 2f  , -f           },
         { -f*Exp[I q] , 0  , -f  , 2f           } }
massmat=DiagonalMatrix[{m1,m1,m1,m2}]
mat2=Inverse[massmat].mat1
eigenlist=Table[{x,
            Chop[Eigenvalues[
            mat2/.{f -> 1., m1 -> 0.4, m2 -> 1.0, q-> x}]]},
                {x,-Pi,Pi,Pi/50}]
plotlist=N[Flatten[Table[
                    Map[{#[[1]],Sqrt[#[[2,k]]]}&,
                        eigenlist],{k,4}],1]]
plot1 := ListPlot[plotlist,FrameLabel -> {"q","omega"},
             Frame -> True,Axes -> None,
             FrameTicks ->{{{-Pi,"-Pi"},{-Pi/2,"-Pi/2"},
                           {0,"0"},{Pi/2,"Pi/2"},{Pi,"Pi"}},
                          Automatic}]
eigensys := Thread[Chop[Eigensystem[mat2 /.
                    {f -> 1., m1 -> 0.4, m2 -> 1.0, q->0.0}]]]
```

2.4 Hofstadter Butterfly: hofstadt.c

```c
/***** Hofstadter Butterfly *****/

#include <math.h>
#include <graphics.h>

#define QMAX 400
#define SHIFT 1

main()
{
  int gcd( int , int);
  int ie, n, nold, m, p, q, x;
  double sigma, pi, e, psiold, psi, psinew;
  int gdriver=DETECT,gmode;
  char str[100];

  initgraph(&gdriver,&gmode,"\\tc");
  settextstyle(1,0,1);setcolor(RED);
  outtextxy(SHIFT+100 ,SHIFT+QMAX+10,"Hofstadter Butterfly");
  settextstyle(0,0,1);setcolor(WHITE);
  outtextxy(SHIFT+100,SHIFT+QMAX+60,"Commands: exit,arbitrary");
```

```
pi=acos(-1.);

/* the eigenvalues for q=2 are known
   and have to be drawn separately */
setviewport(SHIFT,SHIFT,SHIFT+QMAX,SHIFT+QMAX,1);
putpixel(0.5*QMAX,(0.5+sqrt(2.)/4.)*QMAX,WHITE);
putpixel(0.5*QMAX,(0.5-sqrt(2.)/4.)*QMAX,WHITE);

for(q = 4; q < QMAX ; q+=2)
{
  sprintf(str,"q=%d",q);
  setviewport(SHIFT+QMAX+10,SHIFT,SHIFT+QMAX+110,SHIFT+20,1);
  clearviewport();
  outtextxy(1,10,str);
  setviewport(SHIFT,SHIFT,SHIFT+QMAX,SHIFT+QMAX,1);

  for(p = 1; p < q; p+=2)
  {
    if ( gcd(p,q)>1) continue;
    sigma = (2.0*pi*p)/q;
    nold = 0;
    for(ie = 0; ie < QMAX+2 ; ie++)
    {
      e = 8.0*ie/QMAX - 4.0 - 4.0/QMAX ;
      n = 0;
      psiold = 1.0; psi = 2.0*cos(sigma) - e;
      if( psiold*psi < 0.0 ) n++ ;

      for( m = 2; m < q/2; m++ )
      {
        psinew = ( 2.0*cos(sigma*m) - e )*psi - psiold;
        if( psi*psinew < 0.0) n++;
        psiold = psi; psi = psinew;
      }

      psiold = 1.0; psi = 2.0 - e;
      if( psiold*psi < 0.0 ) n++ ;
      psinew = ( 2.0*cos(sigma) - e )*psi - 2.0*psiold;
      if( psi*psinew < 0.0) n++;
      psiold = psi; psi = psinew;

      for( m = 2; m < q/2; m++ )
      {
        psinew = ( 2.0*cos(sigma*m) - e )*psi - psiold;
        if( psi*psinew < 0.0) n++;
        psiold = psi; psi = psinew;
      }
      psinew = ( 2.0*cos(sigma*q/2) - e )*psi - 2.0*psiold;
      if( psi*psinew < 0.0) n++;

      x= p * QMAX / (float)q;
      if(n > nold) putpixel(x,ie,WHITE);
      nold = n;
```

```
      }; /* ie-loop */

   }; /* p-loop */
   if(kbhit()) {str[0]=getch();getch();if(str[0]=='e') break;}
 } /* q-loop */
 closegraph();
} /* main */


int gcd( int a, int b)
{
   if( b==0) return a;
   return gcd(b,a%b);
}
```

2.5 Hubbard Model: hubbard.m

```
Print["\n Hubbard model \n"]
sites=Input[" Number of sites: "] ; Print[""]
particles=Input["Number of particles: "]; Print[""]
spinup=Input["How many of them are spin-up particles? "]; Print[""]
spindown=particles - spinup
tkin=-2.*t*(Sum[N[Cos[2*Pi/sites k]],
                {k,-Floor[(spinup-1)/2],Floor[spinup/2]}]+
            Sum[N[Cos[2*Pi/sites k]],
                {k,-Floor[(spindown-1)/2],Floor[spindown/2]}])
Print[" "]
Print[" The ground state energy for U = 0 is: " ,tkin]
left = Permutations[ Table[ If[ j <= spinup, 1, 0],
                           {j,sites}] ]
right = Permutations[ Table[ If[ j <= spindown, 1, 0],
                            {j,sites}] ]
index = Flatten[ Table[ {left[[i]],right[[j]]},
                {i,Length[left]}, {j,Length[right]} ],1]
end = Length[index]
plus[k_,sigma_][arg_]:= ReplacePart[arg,1,{sigma,k}]
minus[k_,sigma_][arg_]:= ReplacePart[arg,0,{sigma,k}]
sign[k_,sigma_,arg_]  := (-1)^(spinup*(sigma-1))*
                        (-1)^(Sum[ arg[[sigma,j]],{j,k-1}])
cdagger[sites+1,sigma_][any_] := cdagger[1,sigma][any]
c[sites+1,sigma_][any_] := c[1,sigma][any]
cdagger[k_,sigma_][0]:= 0
c[k_,sigma_][0]:= 0
cdagger[k_,sigma_][factor_. z[arg_]]:=
                    factor*(1 - arg[[sigma,k]])*
                    sign[k,sigma,arg]*z[plus[k,sigma][arg]]
c[k_,sigma_][factor_. z[arg_]]:=
               factor*arg[[sigma,k]]*
               sign[k,sigma,arg]*z[minus[k,sigma][arg]]
n[k_,sigma_][0]:= 0
n[k_,sigma_][factor_. z[arg_]]:= factor*arg[[sigma,k]]*z[arg]
scalarproduct[a_,0] := 0
```

```
scalarproduct[a_,b_ + c_]:=scalarproduct[a,b]+
                          scalarproduct[a,c]
scalarproduct[z[arg1_],factor_. z[arg2_]]:=
                          factor* If[arg1==arg2,1,0]
H[vector_] = Expand[
              -t*Sum[cdagger[k,sigma][c[k+1,sigma][vector]] +
                     cdagger[k+1,sigma][c[k,sigma][vector]] ,
                  {k,sites},{sigma,2} ] +
                u*Sum[n[k,1][n[k,2][vector]] ,{k,sites}] ]

Print["\n Calculating the Hamiltonian matrix \n"]
Print[" The dimension of this matrix is ",end,
                                  " x ",end,"."]
h = ( hlist = Table[H[z[index[[j]]]], {j, end}];
   Table[ scalarproduct[ z[index[[i]]], hlist[[j]] ],
          {i,end}, {j,end}] )

Print["\n Calculating the energy eigenvalues."]
state = Chop[Table[Flatten[
                {x,Sort[Eigenvalues[h/.{t -> 1.,u -> x}]]}],
                {x, 0., 6., .1}]]

Print[" \n Commands:"]
menu:=(Print["\n
menu   => This menu \n\n
new    => Restart the program \n\n
plot1 => Multiparticle spectrum as a function of U/t \n\n
plot2 => Double occupancy in the ground state \n
          as a function of U/t \n\n
plot3 => Energy per site in the ground state: Comparison \n
          of the exact result for N -> infinity (dashed) \n
          to the values for finite N"] )
menu

plot1 :=  Show[
  Table[ListPlot[Map[{#[[1]],#[[j]]}&,state],
                Frame->True,Axes->None,
                DisplayFunction -> Identity,
                PlotStyle->{PointSize[0], Thickness[0.0]},
                PlotJoined->True,
                FrameLabel->{"U/t","Energy/t"} ],
        {j,2,end+1}], DisplayFunction -> $DisplayFunction]

g[uu_]:= Chop[Sort[Thread[
            Eigensystem[N[ h/.{t -> 1.0,u -> uu}]]]][[1,2]]]
nnsum = Map[#[[1]].#[[2]]&,index]/sites
md[u_]:= (Abs[g[u]]^2).nnsum
plot2:= (ll = Sort[Eigenvalues[ h /. {t -> 1.0, u -> 1.0}]];
          If[ll[[1]]==ll[[2]],
              text = "Warning: Ground state is degenerate!",
              text = ""];
    Plot[md[u], {u,0.0,6},
          Axes -> None, Frame -> True, FrameLabel->
```

```
            {"U/t","Double occupancy",text,""}]  )

plot3:= (groundstate = Map[{#[[1]],#[[2]]/sites}&,state];
            g1 = ListPlot[groundstate,Frame->True,Axes->None,
                   PlotStyle->PointSize[0],PlotJoined->True,
                   FrameLabel->{"U/t","Energy per site/t"},
                   DisplayFunction -> Identity ];
f[u_,0] = 1/4 ;
f[u_,0.0] = 0.25 ;
f[u_,x_] = BesselJ[0,x]*BesselJ[1,x]/(x*(1+Exp[x u/2]));
plotlist = Table[{u,-4.0*
                  NIntegrate[f[u,x],{x,0,5.52}]},{u,0,6,.2}];
g2 = ListPlot[plotlist,
                PlotJoined -> True,DisplayFunction -> Identity,
                PlotStyle -> Dashing[{0.01,0.01}]];
Show[g1, g2, DisplayFunction -> $DisplayFunction] )
```

3.1 Population Dynamics: logmap.c

```c
/*****  Population dynamics *****/

#include <stdlib.h>
#include <graphics.h>
#include <math.h>

int maxx,maxy;

main()
{
  void channel(double);
  int gdriver=DETECT,gmode;
  char ch,str[100];
  double xit=0.4,rmin=.88,r;
  int x,y=50,i,ir;

  initgraph(&gdriver,&gmode,"\\tc");
  maxx=getmaxx();
  maxy=getmaxy();
  setcolor(RED);
  settextstyle(1,0,1);
  outtextxy(100,maxy-80," Logistic map x=4*r*x*(1-x)");
  settextstyle(0,0,1);setcolor(WHITE);
  outtextxy(50,maxy-30,
    " Commands: lower,higher,refresh,print r,channel,arb.,exit ");
  setviewport(1,1,maxx,maxy-100,1);
  r=1.-(1.-rmin)*y/(maxy-100);

  while(1)
  {
    xit=4.*r*xit*(1.-xit);
    x=xit*maxx;
    putpixel(x,y,WHITE);
    if(kbhit())
```

```
      {
        switch(getch())
        {
          case 'l': if(y<(maxy-100)) y+=1;break;
          case 'h': if(y>1)    y-=1;break;
          default : getch();break;
          case 'e': closegraph(); exit(1);
          case 'c': channel(r);break;
          case 'p': sprintf(str," r= %lf",r);
                    outtextxy(1,y,str);
                    break;
          case 'r':  clearviewport();break;
        } /* switch */
        r=1.-(1.-rmin)*y/(maxy-100);
        for(i=0;i<100;i++)  xit=4.*r*xit*(1.-xit);
      } /* if */
    }/* while */
}/* main */

void channel( double r)
{
  double xit=.4;
  int x,y[1000],i;
  clearviewport();
  for (i=0;i<maxx;i++) y[i]=0;
  line(0,maxy-100,maxx,maxy-100);
  while(!kbhit())
  {
    xit=4.*r*xit*(1.-xit);
    x=xit*maxx;
    y[x]++;
    putpixel(x,maxy-100-y[x],WHITE);
  }
  getch();getch();clearviewport();return;
}
```

3.1 Population Dynamics: logmap.m

```
Print ["\n Logistic Map \n"]
f[x_]=4 r x(1-x)
iterf[n_]:=Nest[f,x,n]
plot1:=Plot[Evaluate[{x,f[x],iterf[2],iterf[4]}/.r->.87],
                {x,0,1}, Frame -> True,
      PlotStyle -> {{Thickness[0.002],GrayLevel[0]},
                    {Thickness[0.002],
                     Dashing[{0.003,0.003}]},
                    {Thickness[0.002],GrayLevel[.6]},
                    {Thickness[0.002],GrayLevel[0]}},
      FrameLabel -> {"x","f[f[f[f[x]]]]"} ]

nf[x_,r_]=N[f[x]]
list1[r_,n_]:=NestList[nf[#,r]&,.65,n]
plot2:= (list=list1[.87,40];
```

```
        g1=ListPlot[list,PlotStyle -> PointSize[0.02],
                DisplayFunction -> Identity ];
        g2=ListPlot[list,PlotJoined->True,
            PlotStyle -> {Thickness[0.002],GrayLevel[.7]},
                FrameLabel -> {"n","x(n)"},
                DisplayFunction -> Identity ];
        Show[g2, g1, Frame -> True,
            DisplayFunction -> $DisplayFunction] )

plot3[r_:0.87464]:=(g1=Plot[nf[x,r],{x,0,1},
                DisplayFunction -> Identity];
        xx=Nest[nf[#,r]&,0.3,500];
        li1=NestList[{nf[#[[1]],r],nf[#[[2]],r]}&,
                {xx,nf[xx,r]},1000];
        li2 = Flatten[Map[{{#[[1]],#[[2]]},
                        {#[[2]],#[[2]]}}&,li1],1];
        li3=Flatten[Map[{{#[[1]],#[[2]]}}&,li1],1];
        g2=ListPlot[li3,PlotRange -> {{0,1},{0,1}},
                PlotStyle -> PointSize[0.015],
                DisplayFunction -> Identity ];
        g3=ListPlot[li2,PlotRange -> {{0,1},{0,1}},
                PlotJoined -> True,
                PlotStyle -> Thickness[0],
                DisplayFunction -> Identity ];
        g4=Graphics[Text[
            StringJoin["r = ",ToString[r]],{0.5,0.05}]];
        g5=Graphics[Line[{{0,0},{1,1}}]];
          Show[g1,g2,g3,g4,g5,PlotRange -> {{0,1},{0,1}},
            AspectRatio -> Automatic,
            Frame -> True,
            DisplayFunction -> $DisplayFunction])

h[x_]=97/25 x(1-x)
hl[n_]:=NestList[h,N[1/3],n]
hl[n_,prec_]:=NestList[h,N[1/3,prec],n]
tab:=Table[{prec,Precision[Last[hl[100,prec]]]},
        {prec,16,100}]
plot4:=ListPlot[tab, Frame -> True,
            PlotStyle -> PointSize[0.01],
            FrameLabel ->
            {"Precision of the calculation",
             "Accuracy of the result"}]

period[1]={c,1}
accuracy=30
maxit=30
period[n_]:=period[n]=Join[period[n-1],
                        correct[period[n-1]]]
correct[list_]:=Block[{sum=0,li=list,l=Length[list]},
                    Do[sum+=li[[i]],{i,2,l}];
                    If[OddQ[sum],li[[1]]=0,li[[1]]=1];
                    li]
g[n_,mu_]:=Block[{x=Sqrt[mu],l=Length[period[n]]},
```

```
                    Do[x=Sqrt[mu+(-1)^(period[n][[i]]) x],
                       {i,1,3,-1}];  x]
fr[n_]:=fr[n]=(find=FindRoot[g[n,mu]==mu,{mu,{15/10,16/10}},
                       AccuracyGoal->accuracy,
                       WorkingPrecision->accuracy,
                       MaxIterations->maxit];
                       mu/.find)
rr[n_]:= rr[n] = (1+Sqrt[1+4*fr[n]])/4
delta[1]:= Print["   The value of n is too small"]
delta[2]:= Print["   The value of n is too small"]
delta[n_]:= delta[n]=(Sqrt[1+4*fr[n-1]]-Sqrt[1+4*fr[n-2]])/
                       (Sqrt[1+4*fr[n]]-Sqrt[1+4*fr[n-1]])
feigenbaum :=
Do[Print["        n = ",n,
          "  Feigenbaum constant = ",delta[n]],{n,3,10}]
```

3.2 Frenkel–Kontorova Model: frenkel.c

```c
/*****  Frenkel-Kontorova Model  *****/
#include <conio.h>
#include <math.h>
#include <stdlib.h>
#include <graphics.h>

main()
{
  int gdriver=DETECT,gmode;
  int xs,ys,color=WHITE,maxx,maxy;
  long nsum=0;
  double k=1.,sig=.4,p=0.5,x=0.55,pi=M_PI,xnew,pnew,h=0.,wind=0.,
         xold,pold;
  char ch,str[1000];

  initgraph(&gdriver,&gmode,"\\tc");
  maxx=getmaxx();
  maxy=getmaxy();
  settextstyle(1,0,1);setcolor(RED);
  outtextxy(150,maxy-80,"Frenkel-Kontorova Model");
  settextstyle(0,0,1);setcolor(WHITE);
  outtextxy(100,maxy-35,
      "Commands: exit, print, new start, clear, arbitrary");
  xold=x; pold=p;
  while(1)
  {
    pnew=p+k/2./pi*sin(2.*pi*x);
    xnew=pnew+x;
    xs=fmod(xnew+100000.,1.)*maxx+1;
    ys=(1.-fmod(pnew+100.,1.))*(maxy-100)+1;
    putpixel(xs,ys,color);
    wind+=xnew-x;
    x=xnew; p=pnew;
    h+=k/4./pi/pi*(1.-cos(2.*pi*x))+(p-sig)*(p-sig)/2.;
    nsum++;
```

```
        if(kbhit())
        { switch(getch())
          {
            case 'e' : closegraph(); exit(1);
            case 'n' : x=xold=(double) rand()/RAND_MAX;
                       p=pold=(double) rand()/RAND_MAX;
                       color=random(getmaxcolor())+1;
                       h=0.;wind=0;nsum=0;break;
            case 'p' : sprintf(str,
                    " Energy= %lf  Winding number= %lf  (x,p)= %lf,%lf",
                          h/nsum,wind/nsum,xold,pold);
                       setviewport(10,maxy-20,maxx,maxy,1);
                       clearviewport();
                       outtextxy(1,1,str);
                       setviewport(1,1,maxx,maxy,1); break;
            case 'c' : setviewport(1,1,maxx,maxy-95,1);
                       clearviewport();
                       setviewport(1,1,maxx,maxy,1);break;
            default: getch();break;
          }/* switch */
        }/* if */
      }/* while */
}/* main */
```

3.2 Frenkel–Kontorova Model: frenkel.m

```
Print[" \n Frenkel-Kontorova Model \n"]
pi = N[Pi]
k=1.
sigma=.4
nmax=1000
t[{x_,p_}] = {x + p + k/(2 pi) Sin[2 pi x],
                p + k/(2 pi) Sin[2 pi x] }
list[x0_,p0_]:= NestList[t,{x0,p0},nmax]
xlist[x0_,p0_]:= Map[First,list[x0,p0]]
tilde[{x_,p_}]:= {Mod[x,1],Mod[p,1]}
listt[x0_,p0_]:= Map[tilde,list[x0,p0]]
plot1[x0_:.06,p0_:.34]:= ListPlot[listt[x0,p0],
                          Frame -> True,  Axes -> None,
                          FrameLabel->{"x","p"},
                          RotateLabel -> False,
                          PlotStyle -> PointSize[0.00]]

plot2[x0_:.06,p0_:.34]:=
     (xl=Map[First, NestList[t,{x0,p0},10]];
      tab=Table[{xl[[m]],k/(2 pi)^2*(1-Cos[2 pi xl[[m]]])},
               {m,11}];
      p1=ListPlot[tab, PlotStyle -> PointSize[0.03],
                      DisplayFunction->Identity];
      p2=Plot[k/(2 pi)^2*(1-Cos[2 pi x]),
              {x,xl[[1]],xl[[11]]},
              DisplayFunction->Identity];
```

```
       Show[p1,p2,DisplayFunction->$DisplayFunction,
                 Frame -> True] )

plot3[x0_:.06,p0_:.34]:=
     (x1=Map[First, NestList[t, {x0,p0}, 99]];
      ListPlot[Mod[x1,1],
                Frame -> True,FrameLabel->{"n","x(n)"},
                PlotStyle -> PointSize[0.013] ])

de[{x_,p_}]:=k/(2 pi)^2*(1-Cos[2 pi x])+0.5*(p - sigma)^2
h[x0_:.0838,p0_]:=(l11=list[x0,p0]; l12 = Map[de,l11] ;
                   Apply[Plus,l12]/Length[l12])
wind[x0_:.0838,p0_]:=(w1=xlist[x0,p0];
                        (w1[[-1]]-w1[[1]])/nmax )
hp0:=Table[{p0,h[.0838,p0]},{p0, .2, .4, .01}]
plot4:= ListPlot[hp0, Frame -> True,
                 FrameLabel->{"p0","energy"},
                 PlotStyle -> PointSize[0.013]]
hx0:=Table[{x,h[x,.336]},{x,.15,.2,.001}]

plot5:=ListPlot[hx0,Frame -> True,Axes -> None,
                FrameLabel->{"x0","energy"},
                PlotStyle -> PointSize[0.013]]

windp0:=Table[{p0,wind[p0]},{p0,0.2,.4,.01}]
plot6:=ListPlot[windp0, Frame -> True,
                FrameLabel->{"p0","winding number"},
                PlotStyle -> PointSize[0.013] ]
```

3.3 Fractal Lattice: sierp.c

```c
/*****  Fractal Lattice  *****/

#include <graphics.h>
#include <stdlib.h>
#include <math.h>

main()
{
  int gdriver=DETECT,gmode;
  struct {int x;int y;} pt={10,10},pw,
                     p[3]={{1,1},{500,30},{200,300}};

  initgraph(&gdriver,&gmode,"\\tc");
  settextstyle(1,0,1);setcolor(RED);
  outtextxy(150,getmaxy()-100,"Sierpinski gasket");
  getch(); setcolor(WHITE);

  while(!kbhit())
  {
    pw=p[random(3)];
    pt.x=(pw.x+pt.x)/2;
    pt.y=(pw.y+pt.y)/2;
```

```
    putpixel(pt.x,pt.y,WHITE);
  }
  getch();getch();
  closegraph();
}
```

3.3 Fractal Lattice: sierp.m

```
Print["\n Generates and Plots a Sierpinski Gasket \n"]
list={{{0.,0.},{.5,N[Sqrt[3/4]]},{1.,0.}}}
mult[d_]:=Block[ {d1,d2,d3},
         d1={d[[1]],(d[[2]]+d[[1]])*.5,(d[[3]]+d[[1]])*.5};
         d2=d1+Table[(d1[[3]]-d1[[1]]),{3}];
         d3=d1+Table[(d1[[2]]-d1[[1]]),{3}];
                 {d1,d2,d3} ]
plot1:= Block[{list2,plotlist},
            list2=Map[mult,list];
            list=Flatten[list2,1];
            plotlist=Map[Polygon,list];
            Show[Graphics[plotlist],
                AspectRatio -> Automatic]]
```

3.4 Neural Network: nn.c and nnf.c

nn.c is available on CD-ROM only.

```
/*****  Neural Network - Reading from a File  *****/

#include <stdlib.h>
#include <math.h>
#include <string.h>
#include <stdio.h>

#define N 10
#define N2 20

float runs=0,correct=0;

main(int argc, char * argv[] )
{
  int neuron[N2],input,i;
  float weight[N2],h,wsum,kappa=1.;
  char ch,str[100];
  FILE * fp;

  if(argc==1)
  {
    printf(" Input file name? ");
    scanf("%s",str);
  }
  else strcpy(str,argv[1]);
  if((fp=fopen(str,"r"))==NULL)
```

```
{
  printf(" File %s not found! ",str);
  exit(1);
}
printf(" File %s is being processed ",str);

for(i=0;i<N2;i++) { neuron[i]=1; weight[i]=(float)i/N; }

while(feof(fp)==NULL)
{
  switch(fgetc(fp))
  {
     case '1' : input=1; runs++;break;
     case '0' : input=-1; runs++;break;
     default  : continue;
  }
  for(h=0.,i=0 ;i<N ;i++)  h+=weight[i]*neuron[i];
  for(wsum=0.,i=0; i<N; i++) wsum+=weight[i]*weight[i];
  if(h*input>0.)  correct++;
  if( h*input < kappa*sqrt(wsum) )
    for(i=0;i<N;i++)   weight[i]+=input*neuron[i]/(float)N;
  for(i=N2-1;i>0;i--) neuron[i]=neuron[i-1];
  neuron[0]=input;
}
if(runs!=0)
printf("\n %6.0f valid inputs",runs);
printf("\n %6.2f %% predicted correctly",correct/runs*100.);
fclose(fp);
getch();
}
```

4.1 Runge–Kutta Method: rungek.m

```
Print["\n Runge-Kutta: Pendulum \n"]
RKStep[f_, y_, yp_, dt_]:=
      Module[{ k1, k2, k3, k4 },
              k1 = dt N[ f /. Thread[y -> yp] ];
              k2 = dt N[ f /. Thread[y -> yp + k1/2] ];
              k3 = dt N[ f /. Thread[y -> yp + k2/2] ];
              k4 = dt N[ f /. Thread[y -> yp + k3] ];
              yp + (k1 + 2 k2 + 2 k3 + k4)/6 ]
RungeK[f_List, y_List, y0_List, {x_, dx_}] :=
 NestList[RKStep[f,y,#,N[dx]]&,N[y0],Round[N[x/dx]] ] /;
              Length[f] == Length[y] == Length[y0]

EulerStep[f_,y_,yp_,h_]:= yp + h N[f /. Thread[y -> yp]]
Euler[f_,y_,y0_,{x_,dx_}]:=
 NestList[EulerStep[f, y, #, N[dx]]&, N[y0], Round[N[x/dx]]
                                   ]
hamilton = p^2/2 - Cos[q]
tmax = 200
dt = 0.1
phi0 = Pi/2
```

```
p0 = 0
r = 0.05
phase:=RungeK[{D[hamilton,p],-D[hamilton,q]},
                {q, p}, {phi0,p0}, {10,dt}]
plot1:=ListPlot[phase,PlotJoined->True,Frame -> True,
                AspectRatio->Automatic,
                FrameLabel -> {"q","p"},
                RotateLabel -> False,
                PlotRange -> {{-2.3,2.3},{-1.85,1.85}}]

phase2:=RungeK[{p,-Sin[q]-r p},
                {q,p},{phi0,p0},{tmax,dt}]
plot2:= ListPlot[phase2,PlotJoined->True, PlotRange -> All,
                AspectRatio->Automatic,Frame -> True,
                FrameLabel -> {"q","p"},
                RotateLabel -> False,
                PlotStyle -> Thickness[0.003] ]

phase3:= Euler[{D[hamilton, p], -D[hamilton, q]},
                {q, p}, {phi0, p0}, {10, dt}]
plot3:= ListPlot[phase3,PlotJoined -> True, Frame -> True,
                AspectRatio -> Automatic,
                FrameLabel -> {"q","p"},
                RotateLabel -> False,
                PlotRange -> {{-2.3,2.3},{-1.85,1.85}}]
```

4.2 Chaotic Pendulum: pendulum.c

```
/*****  Chaotic Pendulum *****/

#define float double
#include <graphics.h>
#include <stdlib.h>
#include <math.h>
#include <stdio.h>
#include "\tc\recipes\nr.h"
#include "\tc\recipes\nrutil.h"
#include "\tc\recipes\nrutil.c"
#include "\tc\recipes\odeint.c"
#include "\tc\recipes\rkqc.c"
#include "\tc\recipes\rk4.c"

double dt=.1,r=.25,a=.5,pi,ysc=3.;
int poincare=0,done=0;
int maxx,maxy,ximage,yimage;
void print(),derivs(double,double*,double*),event(double*);

main()
{
  int gdriver=DETECT,gmode;
  int i,nok,nbad,xold,yold,xnew,ynew;
  double y[3],f[3],t=0.,eps=1e-08;
```

```
initgraph(&gdriver,&gmode,"\\TC");
maxx=getmaxx();
maxy=getmaxy();
ximage=maxx-20;
yimage=maxy-110;
pi=acos(-1.);

y[1]=pi/2.;
y[2]=0.;
print();

while(done==0)
{
  if(kbhit()) event(y);
  if(poincare==1)
  {
    odeint(y,2,t,t+3.*pi,eps,dt,0.,&nok,&nbad,derivs,rkqc);
    xold=fmod(y[1]/2./pi +100.5,1.)*ximage;
    yold=y[2]/ysc*yimage/2+yimage/2;
    rectangle(xold,yold,xold+1,yold+1);
    t=t+3.*pi;
  }
  else
  {
    xold=fmod(y[1]/2./pi +100.5,1.)*ximage;
    yold=y[2]/ysc*yimage/2+yimage/2;
    odeint(y,2,t,t+dt,eps,dt,0.,&nok,&nbad,derivs,rkqc);
    xnew=fmod(y[1]/2./pi +100.5,1.)*ximage;
    ynew=y[2]/ysc*yimage/2+yimage/2;
    if(abs(xnew-xold)<ximage/2) line(xold,yold,xnew,ynew);
    t=t+dt;
  }
}
closegraph();
}

void event(double y[])
{
  switch (getch())
  {
    case 'e': done=1;break;
    case 'c': print();break;
    default : getch();break;
    case 'i': a=a+.01;print();break;
    case 'd': a=a-.01;print();break;
    case '+': ysc=ysc/2.;print();break;
    case '-': ysc=ysc*2.;print();break;
    case 's': y[1]=pi/2.;y[2]=0.;print();break;
    case 't': poincare=!poincare;print();break;
    case 'r': outtextxy(53,42,"a=?");window(13,4,40,4);
              scanf("%lf",&a);print();break;
  }
}
```

```
void derivs (double t,double *y ,double *f)
{
  f[1]=y[2];
  f[2]=-r*y[2]-sin(y[1])+a*cos(2./3.*t);
}

void print(void)
{
  char string[100];
  setviewport(1,1,maxx,maxy,1);
  clearviewport();
  settextstyle(1,0,1); setcolor(RED);
  outtextxy(40,yimage+30,"Forced Pendulum");
  settextstyle(0,0,1); setcolor(WHITE);
  outtextxy(10,yimage+70,
      " Commands: exit,increase a,decrease a,start,read a"
      ",+,-,toggle,clear,arbitrary");
  setcolor(LIGHTGREEN);
  rectangle(9,9,ximage+11,yimage+11);
  setcolor(WHITE);
  sprintf(string,"a=%lf, r=%lf" ,a,r);
  if(poincare==1) sprintf(&string[strlen(string)]," - Poincare");
  outtextxy(320,yimage+40,string);
  setviewport(10,10,ximage+10,yimage+10,1);
}
```

4.3 Stationary States: schroed.c

```
/*****    Schroedinger Equation   *****/

#include <graphics.h>
#include <stdlib.h>
#include <stdio.h>
#include <math.h>

double e=.5,a=0.1;

main()
{
  void image(void), print( double,double,double );
  double k(double),step(double *x,double dx,double*y,double*ym1);
  int gdriver=DETECT,gmode;
  int xs,ysnew,ysold;
  double dx=10./500.0,dxstart=10./500.,y,ym1,yp1,destart=.05,
         de=.05,x,xp1,xm1 ;

  initgraph(&gdriver,&gmode,"\\tc");
  image();

  while(1)
  {
    x=dx/2.;y=ym1=1.;ysold=1;
```

```
    for(xs=1;xs<500;xs++)
    {
      yp1=step(&x,dx,&y,&ym1);
      ysnew=(1.-yp1)*120;
      if(abs(ysnew)>10000) break;
      line(xs-1,ysold,xs,ysnew);
      ysold=ysnew;
    }
    switch (getch())
    {
      case 'e': closegraph();exit(1);
      case 'c': print(e,de,dx);break;
      default : getch();break;
      case '+': e=e+de;print(e,de,dx);break;
      case '-': e=e-de;print(e,de,dx);break;
      case 's': de=destart;dx=dxstart;e=0.01*(int)(100*e);
                print(e,de,dx);
                break;
      case 'l': de=de/10.;print(e,de,dx);break;
      case 'd': dx=dx/2.;print(e,de,dx);break;
    }/* switch */
  }/* while */
}/* main */

double step (double *xa, double dx, double *ya, double *ym1a)
{
  long i,n;
  double k(double);
  double yp1,x,y,ym1,xp1,xm1;

  x=*xa;y=*ya;ym1=*ym1a;
  n=ceil(10./dx/500.);

  for(i=1;i<=n;i++)
  {
    xp1=x+dx;
    xm1=x-dx;
    yp1=(2.*(1.-5./12.*dx*dx*k(x))*y
        -(1.+dx*dx/12.*k(xm1))*ym1)/
        (1.+dx*dx/12.*k(xp1));
    xm1=x;x=xp1;
    ym1=y;y=yp1;
  }
  *xa=x;
  *ya=y;
  *ym1a=ym1;
  return yp1;
}

void print (double e,double de,double dx)
{
  char str[100];
  clearviewport();
```

```
setcolor(GREEN);
line(1,120,500,120);
setcolor(WHITE);
sprintf(str," E= %12.8lf , de=%12.8lf , dx= %12.8lf ",e,de,dx);
outtextxy(10,290,str);
return;
}

void image()
{
  settextstyle (1,0,1);
  setlinestyle(0,0,3);
  setcolor(LIGHTGREEN);
  rectangle(9,9,511,311);setcolor(RED);
  outtextxy(30,350," Schroedinger equation ");
  settextstyle (0,0,1); setcolor(WHITE);
  outtextxy(20,390,
      "Commands: exit,lower(de),+(e),-(e),start(de,dx),dx/2,"
      "clear screen");
  setlinestyle(0,0,1);
  setviewport(10,10,510,310,1);
  setcolor(WHITE);
}

double k(double x)
{
  return (-pow(x,2)-2.*a*pow(x,4)+2.*e);
}
```

4.4 Solitons: soliton.m

```
Print["\n Solitons \n"]
soliton = -2 Sech[x-4t]^2
plot1:= Plot3D[soliton,{x,-5,5},{t,-1,1},PlotPoints->50]

eq=D[soliton,t]-6*soliton*D[soliton,x]+D[soliton,{x,3}]==0
max:= Ceiling[20/dx]
ustart:=Table[-6 Sech[(j-max/2)dx]^2//N,{j,0,max}]
step[u_]:= (Do[uplus[k]=RotateLeft[u,k],{k,3}];
                u+dt*(6u*(uplus[1]-u)/dx -
                       (uplus[3]-3uplus[2]+3uplus[1]-u)/dx^3))
plot2[i_:3]:=(dx=0.05; dt=0.02; upast=ustart; time=0;
                Do[upres=step[upast];
                   upast=upres;
                   Print["Time ",time=time+dt],{i}];
                   xulist=Table[{(j-max/2)*dx,upres[[j]]},
                                {j,max}];
             ListPlot[xulist,PlotJoined->True,PlotRange->All,
                      Frame -> True, Axes -> {True,False},
                      FrameLabel ->{"x","u(x,0.06)"} ] )

firststep[u_]:=(up1=RotateLeft[u]; up2=RotateLeft[up1];
                um1=RotateRight[u]; um2=RotateRight[um1];
```

```
                        u+dt*(3*u*(up1-um1)/dx-
                            (up2-2up1+2um1-um2)/(2dx^3)) )
step2[u_,w_]:= (up1=RotateLeft[u]; up2=RotateLeft[up1];
               um1=RotateRight[u]; um2=RotateRight[um1];
               w+dt(2(um1+u+up1)*(up1-um1)/dx -
                   (up2-2up1+2um1-um2)/dx^3 ) )
init:=(dx = 0.18; dt = .002; upast=uPast=ustart; time = dt;
      dt=dt/10; upres = firststep[upast];
      Do[ufut=step2[upres,upast];upast=upres;upres=ufut,{9}];
      upast=uPast; dt = time; )
plot3[i_:10]:=(If[ Not[NumberQ[dt]]||dt != .002, init];
               If[time==.002,fin=i-1,fin=i];
               Do[ ufut=step2[upres,upast];
               upast=upres;upres=ufut;time=time+dt,{fin}];
               Print["Time ",time];
               xulist = Table[{((j-max/2)*dx,ufut[[j+1]]},
                            {j,0,max}];
               ts = StringJoin["u(x,",ToString[time],")"];
               uu = Interpolation[xulist];
               Plot[uu[x],{x,-10.,10.},PlotRange->All,
                   Frame -> True, Axes -> None,
                   PlotStyle -> Thickness[0.002],
                   FrameLabel ->{"x",ts}] )

u2[x_,t_]=-12(3+4Cosh[2x-8t]+Cosh[4x-64t])/
             (3Cosh[x-28t]+Cosh[3x-36t])^2
plot4:=Plot3D[u2[x,t],{t,-.2,.2},{x,-5,5},PlotPoints->50,
          PlotRange -> {-10,0},
          Shading -> False,
          MeshStyle -> Thickness[0],
          AxesEdge -> {Automatic,Automatic,{-1,1}},
          ViewPoint -> {-1.78095, -2.06202, 2.5},
          AxesLabel -> {"t","x","u"}]

plot5:=ContourPlot[-u2[x,t],{t,-1,1},{x,-10,10},
                PlotPoints->100,
                ContourShading->False,
                PlotRange -> All,
                ContourSmoothing -> 4,
                FrameLabel -> {"t","x"},
                RotateLabel -> False,
                Contours -> {0.1,0.6,1.1,1.6,2.6,4.0,5.6}]

plot6[tt_:0.3]:= (uprime[x_]=D[u2[x,tt],x]; x=-10.0;
        exactlist={{x,u2[x,tt]}}; d=0.07; upl=uprime[x]^2;
        While[x<10., upr=uprime[x+d]^2;
            d = 0.07/Sqrt[1+Max[upl,upr]];
            x = x+d; AppendTo[exactlist,{x,u2[x,tt]}];
            upl=upr ];
        ListPlot[exactlist, PlotRange -> All,
                PlotStyle -> PointSize[0.002],
                Frame -> True, Axes -> None] )
```

4.5 Time-dependent Schrödinger Equation: wave.c

Available on CD-ROM only.

5.1 Random Numbers: random.m

```
Print["\n Random Numbers \n"]
x0 = 1234
 a = 106
 c = 1283
 m = 6075
rndm[x_] = Mod[a*x + c,m]
uniform = NestList[rndm,x0,m+1]/N[m]
triplet = Table[Take[uniform,{n,n+2}],{n,1,m,3}]
unitvectors = Map[(#/Sqrt[#.#])&,triplet]
plot1:=Show[Graphics3D[{PointSize[0.004],
            Map[Point,unitvectors]}],
            ViewPoint -> {2,3,2}]

v0 = {276., 164., 442.}/m  (* vector closest to {0,0,0} *)
(* b1,b2,b3 span unit cell *)
b1 = {-113., 172., 7.}/m
b2 = {173., 113., -172.}/m
b3 = {345., 120., 570.}/m
vp = m*(9*b1+4*b2)+{0,0,100}
plot2 := Show[Graphics3D[{PointSize[0.004],
              Map[Point,triplet]}],
              ViewPoint -> vp]

uni = Table[Random[],{m+2}];
tri = Table[Take[uni,{n,n+2}],{n,m}]
univec = Map[(#/Sqrt[#.#])&,tri]
plot3:=Show[Graphics3D[{PointSize[0.004],
            Map[Point,univec]}],
            ViewPoint->{2,3,2}]
```

5.1 Random Numbers: mzran.c

```
/***** Random Numbers *****/

#define N 1000000

unsigned long x=521288629, y=362436069, z=16163801,
              c=1, n=1131199209;

unsigned long mzran()
{ unsigned long s;

  if(y>x+c) {s=y-(x+c); c=0;}
  else      {s=y-(x+c)-18; c=1;}
  x=y; y=z; z=s; n=69069*n+1013904243;
  return (z+n);
```

```
}
main()
{
  double r=0.;
  long i;
  printf("\n\n 1,000,000 random numbers are generated",
         " and averaged.\n\n");
  for (i=0;i<N; i++)
    r+=mzran()/4294967296.;
  printf ("r= %lf \n",r/(double)N);
  getch();
}
```

5.2 Fractal Aggregates: dla.c

```
/*****  Fractal Aggregates  *****/

#include <graphics.h>
#include <stdlib.h>
#include <math.h>

#define lmax  220
#define rs (rmax+3.)
#define rd (rmax+5.)
#define rkill (100.*rmax)

    char xf[lmax][lmax] ;
    int rx,ry,maxx,maxy,count=0,color=0;
    double rmax=1.,pi;

    void main ()
{
    void occupy(),jump(),aggregate(),circlejump();
    char check();
    int driver=DETECT,mode;
    int i,j;

    initgraph (&driver, &mode, "\\tc");
    maxx=getmaxx();
    maxy=getmaxy();
    setcolor(LIGHTGREEN);
    settextstyle(1,0,1);
    outtextxy(5,maxy-40," Diffusion Limited Aggregation");
    setcolor(WHITE);
    settextstyle(0,0,1);
    outtextxy(maxx/2+30,maxy-30,"Commands: exit, arbitrary");
    randomize();
    pi=acos(-1.);

    for(i=0;i<lmax;i++)
        for(j=0;j<lmax;j++)  xf[i][j]=0;
    xf[lmax/2][lmax/2]=1;
```

```
      occupy();
      jump();

      while(1)
      {
         switch(check())
         {
           case 'k':occupy();jump();break;
           case 'a':aggregate();occupy();jump();break;
           case 'j':jump();break;
           case 'c':circlejump();break;
         }
         if (kbhit())
         switch(getch())
         {
           case 'e': closegraph();exit(1);
           default : getch();break;
         }
    }/* while */
}/* main */

 void jump()
 {
    switch(random(4))
    {
      case 0: rx+=1;break;
      case 1: rx+=-1;break;
      case 2: ry+=1;break;
      case 3: ry+=-1;break;
    }
 }

 void aggregate()
 {
   double x,y;
   xf[rx+lmax/2][ry+lmax/2]=1;
   x=rx;y=ry;
   rmax= max(rmax,sqrt(x*x+y*y));
   if(rmax>lmax/2.-5.) {printf("\7");getch();exit(1);}
   if(count++ %100==0) setcolor(color++ %getmaxcolor() +1);
   circle(4*rx+maxx/2,4*ry+(maxy-30)/2,2);
 }

 void occupy()
 {
   double phi;
   phi=(double)rand()/RAND_MAX*2.*pi;
   rx=rs*sin(phi);
   ry=rs*cos(phi);
 }


 void circlejump()
```

```
{
  double r,x,y,phi;
  phi=(double)rand()/RAND_MAX*2.*pi;
  x=rx; y=ry; r=sqrt(x*x+y*y);
  rx+=(r-rs)*sin(phi);
  ry+=(r-rs)*cos(phi);
}

char check()
{
  double r,x,y;
  x=rx;
  y=ry;
  r=sqrt(x*x+y*y);
  if(r>rkill)    return 'k';
  if(r>=rd) return 'c';
  if(xf[rx+1+lmax/2][ry+lmax/2]
    +xf[rx-1+lmax/2][ry+lmax/2]
    +xf[rx+lmax/2][ry+1+lmax/2]
    +xf[rx+lmax/2][ry-1+lmax/2]>0) return 'a';
  else return 'j';
}
```

5.3 Percolation: perc.c and percgr.c

percgr.c is available on CD-ROM only.

```
/***** Percolation *****/

#include <graphics.h>
#include <conio.h>
#include <stdlib.h>
#include <math.h>

main()
{
  int gdriver=DETECT,gmode;
  double p=.59275;
  int i,j,pr,L=700;

  pr=p*RAND_MAX;
  initgraph(&gdriver,&gmode,"\\tc");
  for(i=0;i<L;i++)
  for(j=0;j<L;j++)
    if(rand()<pr)  putpixel(i,j,WHITE);
  getch();
  closegraph();
}
```

5.4 Polymer Chains: reptatio.c

Available on CD-ROM only.

5.5 Ising Ferromagnet: ising.c

```c
/***** Ising Ferromagnet: Text Mode  *****/

#include <time.h>
#include <stdlib.h>
#include <math.h>
#include <dos.h>
#define L 20
#define VSEG 0xb800

int s[L+2][L+2],bf[3];
double temp;

main()
{
  void setT(double);
  void event(void);
  void frame(int , int);
  char ch;
  int mcs,x,y,e,v;
  clock_t start,end;
  clrscr();
  randomize();
  mcs=0;
  gotoxy(50,1);
  printf("Ising ferromagnet ");
  gotoxy(50,2);
  printf("Monte Carlo simulation");
  gotoxy(50,3);
  printf("System size %d * %d",L,L);
  gotoxy(1,25);
  printf(" Commands: l(ower),h(igher),j(ump),e(xit),arbitrary");
  frame(2*L+3,L+2);

  for(x=0;x<L+2;x++) for(y=0;y<L+2;y++)  s[x][y]=1;
  for(x=1;x<L+1;x++) for(y=1;y<L+1;y++)
  {
    gotoxy(2*(x-1)+3,y+1);
    putch(1);
  }
  getch();
  setT(2.269);
  start=clock();
  while(1)
  {
    if (kbhit()) event();

    for(x=1;x<L+1;x++) for(y=1;y<L+1;y++)
    {
      e=s[x][y]*(s[x-1][y]+s[x+1][y]+s[x][y-1]+s[x][y+1]);
      if( e<0 || rand()<bf[e/2] )
      {
        s[x][y]=-s[x][y];
```

```
            v=2*(x*80+2*(y-1)+2);
            ch=(s[x][y]+1)*15;
            poke(VSEG,v,0xf00|ch);
            /* gotoxy(2*(x-1)+3,y+1);    */
            /* putch(ch);                */
        } /* if */
    } /* for */
    for(x=1;x<L+1;x++)
    {
      s[0][x]   =s[L][x];
      s[L+1][x]=s[1][x];
      s[x][0]   =s[x][L];
      s[x][L+1]=s[x][1];
    }
    mcs++;gotoxy(50,15);printf(" mcs/spin : %d",mcs);
    end=clock();
    gotoxy(50,16);
    printf(" CPU time : %.3e",(end-start)/CLK_TCK/(L*L)/mcs);
  }  /* while */
}/* main */

void setT(double t)
{
  temp=t;
  bf[2]=RAND_MAX*exp(-8./temp);
  bf[1]=RAND_MAX*exp(-4./temp);
  bf[0]=RAND_MAX/2;
  gotoxy(50,10);printf(" Temperature: %.2f",temp);
}

void event()
{
  switch(getch())
  {
    case 'h':setT(temp+=.05);break;
    case 'l':setT(temp-=.05);break;
    case 'j':setT(1.)  ;break;
    default :getch();break;
    case 'e':clrscr();exit(1);
  }
}

void frame(int xmax,int ymax)
{
  int i=0;

  gotoxy(1,1);
  while(i++,i<xmax) putch(205);
  for(i=1;i<ymax-1;i++)
  {
    gotoxy(1,i+1);
    putch(186);
    gotoxy(xmax,i+1);
```

```
      putch(186);
   }
   i=0;
   gotoxy(1,ymax);
   while(i++,i<xmax) putch(205);
   gotoxy(1,1);putch(201);
   gotoxy(xmax,1);putch(187);
   gotoxy(1,ymax);putch(200);
   gotoxy(xmax,ymax);putch(188);
}
```

5.6 Traveling Salesman Problem: travel.c

Available on CD-ROM only.

Index

Onsager, L. 193, 197
Operator 75
– algebra 74
– unitary 149
Optimization
– combinatorial 203
– continuous 201
– discrete 201
– linear 42, 44, 201
Orbit 97, 122
– chaotic 87, 98, 100, 127
– periodic 87, 88
– superstable 84, 85, 90
Order
– magnetic 172
Order parameter 199
Oscillation 138
– damped 126
– longitudinal 59
Oscillator
– anharmonic 48, 131
– harmonic 49
Oscillatory circuit 52

Parallel circuit 54, 58
Parameter set 28
Parameter space 27
Parameter vector 23, 28
ParametricPlot3D 37, 214
Partition function 190
Path length, normalized 207
Pauli exclusion principle 71, 72
Payout table 42, 43, 46
PDF 26
Peak voltage 19, 210
Peierls trick 64
Pendulum 6, 119
– harmonic 9
– nonlinear 122
Perceptron 107, 112
– learning rule 109
Percolation cluster 174
Percolation threshold 172, 175
Period 146, 158, 161
– doubling 127
– relatively prime 161
Permutation 201
Permutations 74, 216
Perturbation 49, 83
Phase 56
Phase difference 52
Phase factor 146
Phase space 9, 11, 120, 122

Phase-space diagram 95
Phase transition 39, 82, 172, 174, 182, 190, 198
Phase velocity 152
Phonon 59, 62
Pixel 86, 103, 134
Plot 8, 141, 213
Plot3D 32, 213
PlotRange 8
PlotVectorField 32, 35
Plus 4
Poincaré, H. 121
Poincaré section 82, 94, 123, 126
Point 159
Pointer 5, 230, 233
Polygon 105
Polymer configuration 183
Polymer dynamics 183
Polymer molecule 102, 182, 186
Polynomial 68
Potential 37, 131
– anharmonic 48, 134
– chemical 199
– commensurate 65
– electric 107
– electrostatic 30
– periodic 93, 97
– quadratic 59
– symmetric 131
– well 42, 121
Power dissipation 58
Power singularity 193
Power spectrum 19
Precision 91
Predictor–corrector method 119
Print 141, 218, 219
printf 5, 226
Probability 43, 145
– density 164
– of presence 47
– overall 148
Probability distribution 154
Procaccia, I. 123
Procedure 4
Product ansatz 47
Pseudo-random number 157, 159, 162
Pulse, rectangular 16

Quadratic form 32
Quadrupole approximation 33
Quadrupole moment 30–32
Quantile 26
Quantum Monte Carlo method 78

Springer
and the
environment

Springer